Fine WoodWorking on Joinery

Fine WoodWorking on Joinery

36 articles selected by
the editors of
Fine Woodworking
magazine

The Taunton Press

Cover photo by Joe Felzman

First printing: January 1985
International Standard Book Number: 0-918804-25-6
Library of Congress Catalog Card Number: 84-052096
Printed in the United States of America

A FINE WOODWORKING Book

FINE WOODWORKING® is a trademark of The Taunton Press, Inc.,
registered in the U.S. Patent and Trademark Office.

The Taunton Press, Inc.
Box 355
Newtown, Connecticut 06470

Contents

1   Introduction
2   On Precision in Joinery
6   Mortise and Tenon
12   Mortise and Tenon by Hand
18   Japanese Sliding Doors
27   Locking the Joint
30   The Haunched Mortise and Tenon
32   Louvered Doors
34   Mortise and Tenon by Machine
39   Slip Joints on the Radial-Arm Saw
45   Routing Mortises
48   The Frame and Panel
52   Radial-Arm Raised Panels
54   The Scribed Joint
56   Entry Doors
60   Paneled Doors and Walls
65   The Right Way to Hang a Door
66   Tambour Kitchen Cabinets
67   Questions & Answers
68   Solid Wood Doors
70   Joinery along Curved Lines
74   Relying on the Router
78   Curved Slot-Mortise and Tenon
82   Decorative Joinery
86   Three Decorative Joints
89   Template Dovetails
92   Gluing Up
96   Which Glue Do You Use?
99   Why Glue Joints Fail
100   Glues for Woodworking
103   Dovetail Jigs
106   Tablesawing Dovetails
108   Curved Dovetails
110   The Butterfly Joint
112   Bandsawn Dovetails
113   A Two-Way Hinge
114   Knockdown Furniture
118   Pole-and-Wire Joinery
120   Index

Introduction

Wood as it comes off the tree is generally straight and narrow: sticks and boards. The things we want to make with wood are often broad and three-dimensional: frames, doors, stands and boxes. Thus the art of joinery—that is, how to fit pieces of wood together strongly, as if the tree had grown and branched to our desires.

Joinery is an ancient craft. Egyptian tombs have divulged images of craftsmen beating out mortises with mallet and chisel, images that are uncanny in their familiarity. Hand-woodworkers do it no differently today. At the same time, every new woodworking machine seems instantly to generate a new approach to joinery, to augment or replace a hand-tool technique.

When we join two sticks to make a corner and add two more to make a rectangle, we'll probably use some variation on the mortise and tenon joint. This basic frame joint, in all its myriad forms, is the main subject of *Joinery*, though basic dovetailing and decorative joints are included as well. In 36 articles reprinted from the first nine years of *Fine Woodworking* magazine, authors who are also craftsmen tell exactly how they choose, make and use these joints, and how you can do so too.

When we wrap four frames or four boards around a boxy volume, we're more likely to need the dovetail or some other carcase joint. Case joinery is another large subject, and a companion book in this series, *Boxes, Carcases and Drawers*, is devoted entirely to it.

John Kelsey, editor

On Precision in Joinery

How close is close enough?

by Allan J. Boardman

An exemplar of precise joinery, author's full-blind finger-joined music box is 4½ in. on a side. Carcase is flame-figured butternut, dovetailed drawers are English beech with rosewood pulls.

Comparing a machine tool such as a lathe for shaping metal with its counterpart for working wood suggests that entirely different methods and standards are normally applied when operating on these two dissimilar materials. The differences are obvious—finely graduated scales and dials festoon the metal lathe, while the wood lathe probably has no measuring scales at all. What may not be obvious is the fact that woodworkers nonetheless do approach tolerances that might seem appropriate only to metal. The flexibility and compressibility of wood, the acceptability of fillers, moldings and bulk-strength adhesives, the dynamic movement of the material and the omnipresence of shoddy commercial products all contribute to the belief that "precision" is not a word in the woodworker's vocabulary. However, a close look at a truly fine piece of cabinetry will reveal some surprising facts about the dimensional tolerances inherent in its joinery.

Consider the miter joint connecting two adjacent members of a frame made from 3-in. wide stock (figure 1). If the miter were tight at one end and open, say, ⅟₆₄ in. (0.016 in.) at the other, the joint would be quite unacceptable. The frame would be weak, since most adhesives work best in films far thinner than ⅟₆₄ in. Even an untrained eye could easily detect the mismatch, and filler could not disguise it.

Most shops have lots of clamps, and all too often they are used in abundance to bend or press a joint closed while the glue dries. The result may well be a tight joint, but the structure is liable to be distorted—warped, bowed or out-of-square. This distortion may cause extra work in fitting for doors or drawers, perhaps some unanticipated cosmetic repairs, or it may even be uncorrectable and quite obvious in the finished product. And regardless of how well one compensates, the assembly will retain residual stress after the clamps are released. Built-in stress will work against the adhesive for a long time, causing the joints to creep and the dimensions to change. Stress can burst open an otherwise strong joint months or even years later. Improperly seasoned wood and changing humidity, although usually contributory, are sometimes blamed for joint failure when the real problem is faulty joinery initially hidden by clamping pressure. In first-class work, there is no substitute for joints that fit properly.

In figure 1, note that the angle of the tapered space in the miter joint is less than a fifth of a degree. The tolerance in a good miter might be ⅟₁₀ of that, or barely 1 minute of arc. With such a fit, the open end of the tapered gap would be less than 0.002 in., or about half the thickness of a piece of paper. This, in most cases, would be acceptable from the standpoint of strength and appearance. But measurement and tolerances in thousandths of an inch and minutes of arc sound like the language of machinists, not woodworkers. After all, many of our measuring devices are themselves made of this changeable stuff, wood. The protractor scale on a woodworking machine goes no finer than one degree—minutes of arc,

never. Parallax caused by the distance between pointer arrow and protractor ensures significant error, depending on where you hold your head. Does no one expect a woodworker to hold to a small fraction of a degree, except perhaps at 90° and 45°, where some machines have detents?

So it is with lineal dimensions too. For the seasoned worker, tricks, techniques, experience and feel (not mutually exclusive terms) compensate for the limitations of the equipment. But to the beginner, the not-quite-square square, the coarse graduations of scales and protractors, the machine's structural flexibility where rigidity is desired, all subtly suggest that only this crude level of accuracy is to be expected. Worse, because of careless use of words like "precision," "accurate," "professional" and "heavy duty" in advertising, the novice comes to believe that plus or minus a thirty-second is precise or that the machine by itself guarantees precision. Consequently, beginners may set personal standards for quality far lower than they should and progress far too slowly in the acquisition of those skills and techniques needed to overcome tool limitations.

Tool quality, measuring and marking—The limitations of our tools are not all bad, once recognized and understood. If a manufacturer were to add the weight, rigidity and precision some of us dream about, the cost of tools would rapidly become prohibitive. Also, because of the properties of wood, some of this extra precision would be wasted: The skilled maker would still have to compensate for the peculiarities of each species and piece.

Some tool limitations may require us to take lighter cuts, and they may inhibit some design options or demand greater skill, but by one means or another, we live with the available tools. Nonetheless, the first thing we must do is correct what can be corrected. For example, to true a metal framing square, draw a diagonal from the inner corner to the outer corner. With a hammer, gently tap a center-punch in several places along the inner third of the line to open it, or along the outer third of the line to close it, checking the square after you make each adjustment. Cabinetmaker's squares with a metal blade and wooden stock can be filed true. Bench planes require all sorts of fine tuning before one can realize their full potential.

Leaving aside heavy-duty production machinery, one should not take for granted the implied precision or quality of tools. If you have the time and patience (and the indulgence of the shopkeeper) to examine and compare all of the squares, planes or chisels in stock, you may find one that is better than the others. The common test for a square, for example, is to mark a line on wood or paper taped to the counter, then flop the square over to see if the blade lines up with the line (figure 2). Any discrepancy is double the inaccuracy of the square. But realize that you will have to spend time on most tools to make them right.

So how do you make them right? Against what do you check for square? There is no way around it: Every shop needs a reliable standard for straight, square and flat. A quality machinist's combination square is a good investment because it provides a reliable 12-in. straightedge, an accurate square and a 45° reference. A 3-ft. metal straightedge is useful and is available at some woodworker and most machinist supply houses. One can also buy a strip of flat tool steel and have a machine shop grind it true. The top of a quality table saw should be flat enough to serve as the reference surface, but it is best to check this if possible by removing it, toting it to a machine shop for measurement and, if necessary, having it ground. Other flat references are granite surface plates and slabs of heavy plate glass or marble, which are generally quite flat but must also be checked. The rule of thumb is that these shop standards should be five to ten times better than anything you are likely to check with them. It is also desirable to have at least one fairly large bench surface be rather flat, say within 1/64 in. over a two or three-foot square, for layout work. This can be prepared with a jointer plane and checked with your reference flat, by rubbing one surface against the other through carbon paper. If your reference surface is not easily moved and inverted, use winding sticks instead (figure 3).

An accurate ruler or scale is also important. Simply because a stick or tape is marked in inches and fractions, it does not follow that the marks are where they should be. Some steel tapes are off as much as 1/8 in. in 10 ft. The machinist's combination square will provide a reliable 1-ft. scale against which others in the shop can be calibrated. The graduations are generally fine and deeply engraved for long life.

These points are about absolute accuracy. More important, most of the time, is relative accuracy. Once the dimensions of a given piece are quite close, the requirement for fit outranks the requirement for hitting the exact dimension on the nose. Consider cutting the four moldings for a picture frame. First, the pieces must be near to the desired length. Second, each piece must be the same length as its opposite, and third, after mitering, the corner joints must be tight (figure 4). Because the molding might not be perfectly true or straight, we trim the miter to fit, and as a result the mating surfaces may be a fraction of a degree off the nominal 45°, or one of the sides may be a deliberate but imperceptible fraction shorter than its opposite. A tiny variation in dimension cannot be observed, whereas an open joint will always be visible and weak. At the stage of final fitting, the ruler or gauge becomes a superfluous intermediary, an unwanted source of error.

This notion of dimension giving way to fit is not radical. It is like the intuitive procedure we use when setting a tool or machine whose protractor or scale has only coarse graduations. We guess at a setting someplace between two markings and then, ignoring the actual number of degrees or thousandths, we make small adjustments by trial and error, perhaps with a piece of scrap, until the fit is just right.

Marking can be done with a sharp pencil, but when the position of the mark impacts final fit, a marking knife should be used. Not only will the line be narrower and therefore better define the position, but a knife will lie much closer to the

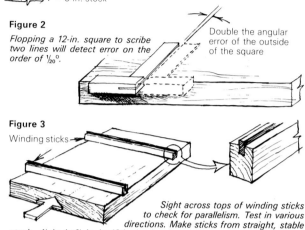

Figure 1

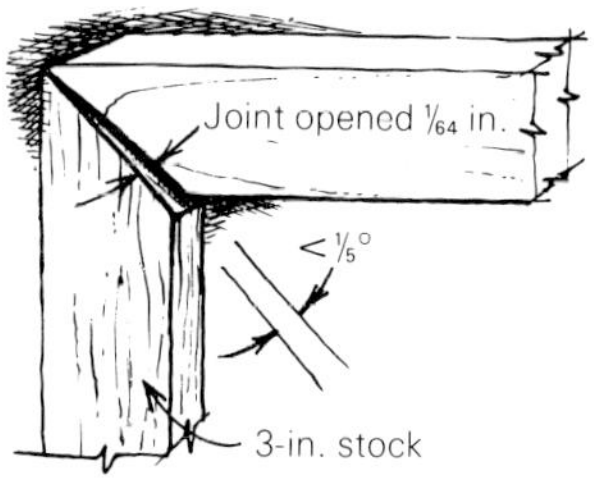

In a frame made of 3-in. stock, a total error of 1/5° in cutting the miters will cause a 1/64-in. gap in the joint. This could result from an error of only 1/10° in setting the saw or in using the miter box, or from warped wood, or even from a tiny chip lodged between the fence and the work. To avoid the error, woodworkers cannot rely on the gross measurement that machine scales provide.

Joint opened 1/64 in.
< 1/5°
3-in. stock

Figure 2

Flopping a 12-in. square to scribe two lines will detect error on the order of 1/20°.

Double the angular error of the outside of the square

Figure 3

Winding sticks

Sight across tops of winding sticks to check for parallelism. Test in various directions. Make sticks from straight, stable stock — 1/2 in. by 3/4 in. by 18 in. is a handy size. Fancy version has insert of light wood in one stick, dark in the other, for better visibility. Well-made sticks used carefully can find 1/10° of error.

Figure 4

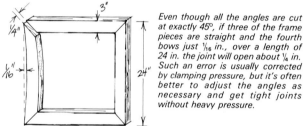

Even though all the angles are cut at exactly 45°, if three of the frame pieces are straight and the fourth bows just 1/16 in., over a length of 24 in. the joint will open about 1/4 in. Such an error is usually corrected by clamping pressure, but it's often better to adjust the angles as necessary and get tight joints without heavy pressure.

3"
1/4"
1/16"
24"

gauge, square or piece used in marking. Furthermore, the mark, being a physical incision in the wood, can often be used to position a chisel for the next operation. A typical example would be marking the shoulder line on a tenon to be cut with hand tools. The knife cut serves simultaneously to locate the shoulder edge, neatly sever the surface grain, and guide a chisel to create a starter groove for the tenon saw.

Cutting to the line—So much for measuring; the marked piece must now be cut. Precision in cutting is the exclusive domain of neither hand nor power tools. I say this despite diehard traditionalists who would argue that truly fine work can be done only by hand, and despite power-tool proponents who believe a plane is what you'd be forced to use if you couldn't afford a machine. There is seldom one best way. A proper table-saw setup would save time if a number of identical tenons were to be cut. The hand-tool method might be best for only one joint, if several different pieces are required or if the shoulder is not perpendicular to the rail.

Often, a combination of hand and power tool methods offers optimum results, taking advantage of the best characteristics of each tool. Suppose tenons are cut at each end of a stile, but for some reason the distance between the tenon shoulders is just a bit fat. (This can usually be avoided by checking with scrap before cutting the work itself.) Moving the table-saw fence a controlled 1/64 in. is tricky. And unless it has just been sharpened, a circular-saw blade is not too effective in trimming off the merest hair. The wood may burn or the blade may deflect, leaving a cocked, charred shoulder.

It is undeniable that power tools save time and physical exertion, but some cuts in precision joinery, such as shaving off that minute error, are clearly better performed with hand edge tools—planes in particular. In practiced hands a shoulder plane can trim that miscut tenon down to size in seconds. End-grain shavings as thin as 0.002 in. can be produced, enabling the scribed line to be approached under the watchful eye of the maker (figure 5). It makes little difference if the waste to be removed is straight or tapered. Because of these factors, it is often advisable when using machines to leave a little margin for hand trimming.

Planes do not cut like most power tools, virtually all of

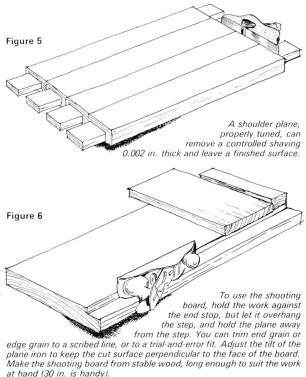

Figure 5

A shoulder plane,
properly tuned, can
remove a controlled shaving
0.002 in. thick and leave a finished surface.

Figure 6

To use the shooting
board, hold the work against
the end stop, but let it overhang
the step, and hold the plane away
from the step. You can trim end grain or
edge grain to a scribed line, or to a trial-and-error fit. Adjust the tilt of the
plane iron to keep the cut surface perpendicular to the face of the board.
Make the shooting board from stable wood, long enough to suit the work
at hand (30 in. is handy).

Figure 7

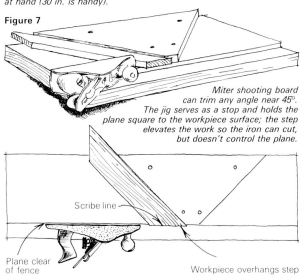

Miter shooting board
can trim any angle near 45°.
The jig serves as a stop and holds the
plane square to the workpiece surface; the step
elevates the work so the iron can cut,
but doesn't control the plane.

Scribe line

Plane clear
of fence

Workpiece overhangs step

which cut intermittently, pounding on the wood fibers and doing inescapable damage on every cut surface. Machine planes, power saws, and even sandpaper produce dust, tearing, burnishing, scratching, and so on. The unique action of the plane with or across the grain, however, severs the wood fibers cleanly in a continuous, not intermittent, motion. The finished surface in many cases cannot be improved upon. The damaging forces involved in parting the waste from the work are absorbed by the shaving as it breaks or curls. The planed surface, except where the wood grain is particularly cantankerous, shows no evidence of the trauma. Even more important for precision is the fact that the plane can leave a good surface after each pass. The perfect fit can be approached by increments and when achieved, no further clean-up is required. In many routine joinery operations, this objective is far more difficult to reach with sandpaper, files, saws or routers. And as a bonus, all this control comes with no great sacrifice in speed. A plane stroke takes only a couple of seconds.

Paring with chisels and other edge tools offers similar possibilities for working toward precise joints, particularly where the geometry prohibits using a plane. Here the control afforded rather automatically by the plane must be provided by the craftsman. However, the principle is the same—taking off just as much waste as desired, exactly where desired, and leaving a clean surface after each cut.

Jigs—The criteria I apply to virtually all joints are first, in hidden interfaces (the tenon in the mortise) there should be only enough clearance for a thin glueline; and second, visible interfaces (miters, for example) should appear tight with only light clamping. As you approach the final fit a shaving at a time, you quickly discover the need for devices that help keep the hand tools perpendicular, free from wobble, or otherwise aligned. Jigs and fixtures do not guarantee precision, but they can reduce the degree of freedom the tool has so the craftsman can exercise greater control toward getting the fit.

A jig of continuing use is the shooting board—nothing more than a flat piece of stable wood with a step at the edge and a stop near one end (figure 6). With it, one can simultaneously plane an end or edge of a piece exactly to a scribed line, straight and perpendicular to the surface. Using this same jig and a little blocking or intentional tilt of the plane blade, angles other than 90° can easily be cut for coopered joints or simply to compensate for some special condition.

Other jigs in this same family include several versions of the bench hook, and the miter shooting board—the solution to the problem that began this essay, of how to adjust a miter angle by a fraction of a degree (figure 7). In use, the plane is laid on its side on the ledge while the work is held against the 45° stop. If the plane body is out of square (it usually is), the mitered surface will not be perpendicular to the face of the piece. This can be corrected to some extent by adjusting the tilt of the plane iron, by shimming the work, or maybe 91° is really desired. The 45° angle (or 44° or 46°20') is not a result of holding the plane firmly against the step in the fixture while pressing the work against the fence. Of course, it could be if the jig were made exactly at the angle desired, but that is too restrictive a use of the shooting board. Rather, one holds the work against the stop but overhung, and the plane sole away from the step. One then planes either to a scribed line or by trial and error to a perfect fit with the mating piece.

In contrast to such "permanent" devices, many simple jigs

can be made for short-term use. The usual reasons such jigs fall into the disposable category are that they get worn or damaged in use or are special in nature or dimensions. Consider cutting dadoes by hand in the two vertical sides of a bookcase (figure 8). A useful multipurpose jig fashioned from two pieces of wood not only simplifies the operation but also facilitates precision. In appearance, the jig is nothing more than a clumsy-looking square, the long leg reaching across the workpiece, the short leg attached accurately at right angles. The width of the members should ensure stiffness and rigidity. The thickness of the short leg should be a trifle less than the workpiece thickness so as not to interfere with clamping. The thickness of the long leg must be sufficient to keep the backsaw perpendicular, but it can also be such that when the saw back hits the jig, the cut is at the desired depth. The jig is clamped to the workpiece and at the one setting serves as a straightedge for scribing, a control for chiseling out a starter groove for the backsaw, a fixture for holding the saw upright, and a depth stop. Two such sequences per dado, followed by cleaning out the waste with chisel or router plane, leave an exceptionally clean joint the width of which can exactly fit the thickness of the shelf. With this method it matters little that the shelves vary in thickness from one to another, or that the dado head on your power saw cuts only in fixed increments that don't match your wood. Notice that in this example since all the scribing and sawing are done on the waste side, both long edges of the jig are used and so must be parallel. Obviously, the same basic technique can be adapted to other and more complicated joints—stopped dadoes, rabbets and dadoes, tapered dovetails and so on.

Precison is relative—In woodworking there is a scale of precision demanded by the nature of each project, from rough to finish carpentry ascending through built-ins to fine cabinetry, furniture-making and ultrafine craft objects like view cameras. Tolerances might range from plus or minus an eighth to one or two thousandths. In addition, we must superimpose a scale of functional tolerances that takes into account the size of the object and the wood's probable movement in response to changes in temperature and humidity. The "precision fit" of a drawer in a fine chest incorporates a neat but wider clearance gap than one would find around the drawer of an equally well-made jewelry box.

Finally, one should not neglect the many different design options that shift the need for one kind of precision to another, or eliminate the need altogether. The results can be quite acceptable and are normally found in abundance on commercial work. Take, for example, the use of a solid nosing around the top of a veneered cabinet (figure 9). To blend the grain of the solid piece with the veneered panel and to join it flush without damaging the thin and delicate veneer would involve considerable skill and risk. This requirement can be virtually eliminated by accentuating the seam instead of hiding it, with a routed or scratched groove used as a design feature. Likewise, moldings can effectively mask imprecise joinery, and overlapping fronts can conceal uneven clearance around cabinet drawers. With design skill, such techniques can permit production shortcuts. Often, they are the best choice in purely design terms, and the fact that less precision is required becomes a bonus.

The characteristics that denote precise woodworking are not limited to joint accuracy and fit. They also include grain

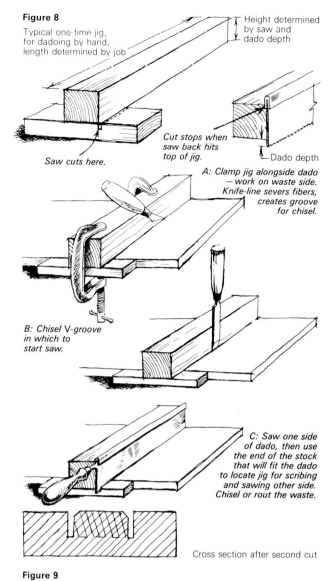

Figure 8

Typical one-time jig, for dadoing by hand, length determined by job

Height determined by saw and dado depth

Saw cuts here.

Cut stops when saw back hits top of jig.

Dado depth

A: Clamp jig alongside dado — work on waste side. Knife-line severs fibers, creates groove for chisel.

B: Chisel V-groove in which to start saw.

C: Saw one side of dado, then use the end of the stock that will fit the dado to locate jig for scribing and sawing other side. Chisel or rout the waste.

Cross section after second cut

Figure 9

Deliberately accenting a joint may be better than trying to hide it.

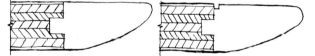

Planing, sanding or scraping solid banding flush may damage veneer.

Scratched or sawn groove highlights joint, reducing need for precision.

and color matching, uniformity of detail and symmetry when intended, clean pre-finish surface preparation, crisp installation of inlays and fittings, minimum use of fillers, and so on.

Precision in joinery is neither for everyone nor for every project. It can be an objective or an attitude that adds pleasure to the craft and quality to the work. It can, on the other hand, become an obsession that goes beyond common sense to the point of inconsistency with the nature of wood itself. But it seems far better for a woodworker to understand the options, recognize that certain skills and techniques can be invented or learned, know what is possible to accomplish, and then exercise free choice, rather than have his or her standards derive from crude scale markings and constant exposure to mediocre work. □

Mortise and Tenon

There are innumerable variations of this basic joint

by Tage Frid

Furniture construction is broken down into two categories — frame and casegood. Casegood construction uses joints such as dovetails, finger joints, spline miters, rabbets and the like. Frame construction depends on the mortise and tenon joint and is usually used in tables, chairs, paneled doors, windows, etc. There are a great many variations of the mortise and tenon joint, and the task of the cabinetmaker is to know which variation to choose for a particular application, and why, and then how to make it quickly and well.

The mortise and tenon is probably the oldest and certainly the most essential joint in woodworking. An Egyptian sarcophagus now in the British Museum was framed with mortise and tenon joints at least five thousand years ago. During the Middle Ages, the development of the mortise and tenon permitted the framing of chests. The elaborate variations of paneling led finally to a distinction between the two crafts of carpentry and cabinetmaking. In house construction the use of the mortise and tenon has quite disappeared. We no longer have the skill or the patience, nor can we afford the mortise and tenon for the framing of a house. Perhaps we do not expect our houses to endure for much more than a few generations. But we do still find esthetic and practical satisfaction in a well-constructed piece of furniture.

The strength of the mortise and tenon joint depends entirely on the interplay between the cheek and shoulder of

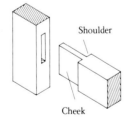

Shoulder

Cheek

the tenon, which is the projecting part of the joint. One can imagine two crossed boards glued together. Despite the holding power of the glue, they can be twisted apart relatively easily.

But connect them as a lap joint, and the strength is in-

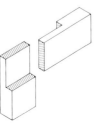

creased greatly because of the stop-action of the shoulders. Now double the surface area of the glue by making a slip joint

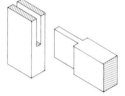

— a form of mortise and tenon — and we have an extremely strong joint that is easy to make and requires minimum tools.

The disadvantage of the slip joint is that not only do we have to clamp the tenon shoulder tightly against the mortise, (as in all mortise and tenon assembly), but we must use a second clamp to make sure the cheeks are glued to the mortise sides. Moreover, the tenon is completely exposed.

We get around these drawbacks by changing the slip to a

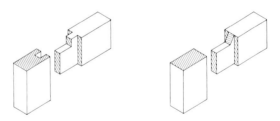

haunched mortise, or to a mitered haunched mortise where the tenon is completely hidden.

When designing a mortise and tenon joint, one should aim for the maximum glue surface. A tenon of about one-third the thickness of the stock is usually the best choice. If the tenon is thicker, the mortise sides become too thin; if the tenon is thinner, it becomes too weak. (But sometimes in table construction, where the leg is much thicker than the aprons, the aprons may have tenons half or more the apron thickness.)

Four shoulders should never be used unless absolutely necessary. The joint becomes more difficult to fit because all

four shoulders must be precisely located in the same plane. Also, glue surface is lost. On the other hand, if the design calls for carving and material will be removed around the joint, four shoulders ensure that the joint will not be revealed.

If the design calls for round corners it is advisable to glue a block on, or to have the mortise stock wider. These provisions

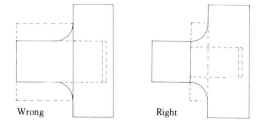

Wrong   Right

prevent problems with the end grain which will break and crumble, especially if carved.

There are two different ways to make a round corner in a frame. The left one is used if the inside corner is going to be

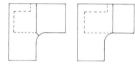

carved or shaped for a molding, and the right one is fine if the edge will be left straight, because then you don't have to worry about carving into the joint.

When a tenon is very wide, haunches should be put in at either end. A wide tenon is more difficult to glue as it

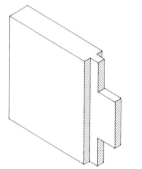

requires extra clamps for gluing the cheeks. But the haunches are necessary to keep the wood from twisting.

When a tenon is very narrow, the temptation is to run the tenon across the grain. But this should never be done because

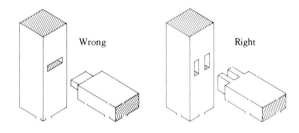

Wrong   Right

then the cheeks glue into end grain which is not a glue surface. The way to fasten narrow tenons is to use double (or triple) tenons, running the mortises in the direction of the long grain to provide good glue surface.

Wedges are used to strengthen the joint. When the tenon is cut to receive the wedge be sure to drill a small hole at the

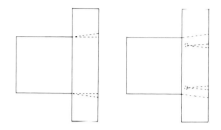

base of the saw cut to prevent cracking. When hammering in wedges in a through tenon, be sure to hammer evenly on each wedge so as not to force one half or the other too far which could result in splitting. If the tenon is to be hidden, use this method.

If a mortise and tenon is to be disassembled, a loose wedge is used. The wedge could be substituted with a wedged dowel

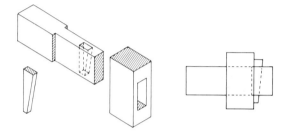

for the same effect. If the piece which receives the wedge is too thin, the two shoulders could be placed on the top and bottom instead of the sides.

In a chair, the back is usually one to two inches narrower than the front. This is done more for appearance than for any other reason. This requires the sides to angle into the back.

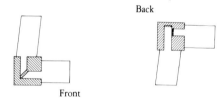

Back

Front

Usually the angle is made in the tenon, because it is easier than angling the mortise. Of course there is a limit to how much the tenon can be angled, but as long as some long grain reaches the full length of the tenon, it is safe.

A variation of the slip joint is used where a third or fourth leg is necessary, as in a sofa. This is also used where a table

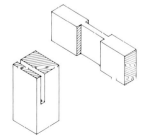

apron is joined to the legs if the table apron is bricklayed round or oval as in a Hepplewhite table.

There are several ways to make a mitered mortise and tenon. Often a spline is used, as it is easier to cut. Sometimes

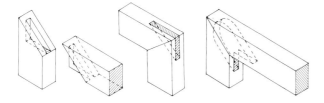

a spline is used purely for visual effect. The spline can also be hidden.

If a tenon should break, a spline can be inserted. The same method is often substituted for a mortise and tenon.

Although it is not as strong, the spline is in most cases sufficient, and is again much easier to make.

There are many other variations of the mortise and tenon joint but virtually all depend on the cheek and shoulder action for their strength. Similarly, the technique used in making these joints is basically the same.

In making mortise and tenon joints, I find it easier and quicker to use hand tools, unless there are so many joints that power tools turn out to be quicker. But this is rarely the case because power tools — whether I'm using a saw, or a router or a drill press with a mortising bit — do take time to set up for the particular job. But even if you plan to use power tools, it's best to learn to do them by hand, so that you understand what you're trying to do with the power tools.

The first step in making the basic two-shoulder joint is to mark both pieces to keep the orientation right. Then I outline

the tenon piece on the mortise piece, but I use a square to put lines just inside (less than 1/16-inch) those marks, because the mortise should be made slightly smaller to allow for subsequent sanding of the tenon.

I pick a drill or bit about 1/3 the thickness of the tenon board. If this size is between bit sizes, I use the next larger one. Although it isn't absolutely necessary, I recommend using a doweling jig to guide your bit while boring the mortise. You'll end up with straight and even sides. Make the

two outside holes first, then the holes in between.

Stop work on the mortise at this point and transfer its dimensions to the tenon board. First measure the depth of the mortise with a rule and make the tenon 1/8-inch shorter

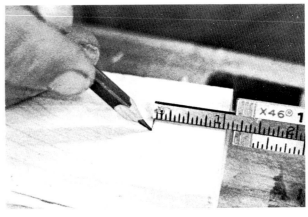

to allow for excess glue. I use a square and a scribe to draw this depth line around all four sides of the tenon board. This marks where the shoulders will go. Don't use a pencil be-

cause its line is too wide and the shoulder must be cut with great accuracy.

Then take a marking gauge and adjust it so its point just touches the nearer side of the holes bored for the mortise. Transfer this measurement to the ends and two sides of the

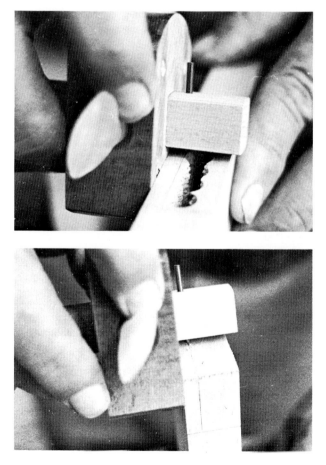

tenon. Then do the same for the other side of the mortise (but continue using the same reference surfaces).

You are now ready to cut the cheeks of the tenon. I use a frame saw for this (as I use for almost all hand sawing because it's the fastest and best saw there is) but if you don't have one, use a dovetail or back saw. The thinner the blade, the easier it will be to make accurate cuts.

The trick to cutting accurate cheeks is to cut the back line and part of the top first; then turn the board around and cut the rest of the top and the front lines. That way you don't have to worry about following two lines at once. When cutting the front line, the saw blade will be automatically

guided at the back by the kerf you made before. You'll also get a little more accuracy in this guiding process if you use a slightly thinner blade for cutting the back lines than you do for cutting the front.

In any event, when sawing the cheeks, "split" the line on the waste side. The tenon cheeks must fit just right. If they're too tight they may split the mortise piece; if too loose, the glue joint may come apart under strain. Furthermore, the surface over a mortise that holds a loose tenon will in time become concave as the wood dries.

After the cheeks are sawed, it's time to saw the shoulders. One trick I've found helpful to improve the accuracy (since the shoulders must be perfectly aligned) is to make in effect a small or mini-shoulder for the saw to lean against. Take one corner of a flat chisel and deepen the shoulder line by

drawing the chisel along it. Then take a second cut at an angle to create half a "vee". You can then use this notch as a guide for your dovetail or frame saw. Finish sawing the shoulders and use a flat chisel to clean up the cheeks, and then round

off the tenon corners slightly for easy insertion. Then sand the edge of the tenon so it will fit into the slightly shorter mortise.

Now finish making your mortise. Take a small chisel and mallet to square off the corners, and a wide chisel (but no mallet) to flatten out the sides. Sand the outside edge of the

mortise piece as you did the tenon sides and you're ready to try the fit. You should be able to push it in by hand with the weight of your body. If you need to hammer it in, it's too tight and you should shave some material off the tenon

assembled, make a clean saw cut along the shoulder line, making sure not to cut into the mortised piece at all. Do the same for the other shoulder. Don't saw quite completely to the tenon. Instead, finish the cuts with a chisel after the joint is disassembled. If you're making a frame and notice one of

the shoulders is off after you've dry clamped it, make the shoulder correction cuts to all the shoulders on the same side of the frame, so that after correction, the frame stays square (but one blade width shorter). Of course, if a shoulder is really off, you may need to go through the correction process twice.

cheeks because that's the easier piece to correct. If the tenon is too loose, you can glue strips of veneer to the cheeks.

If after fitting, the shoulders are slightly off as illustrated here, there's a trick you can use to align them. With the joint

On complicated pieces where the joints may come in at odd angles, I sometimes don't worry about precise fitting of the shoulders during the initial cutting process, but rely instead on the correction cuts to get the fit I want.

When gluing a mortise and tenon joint, it is very important to put a moderate amount of glue in the mouth of the mortise, and just a little on the beginning of the tenon cheeks and on the shoulders (as insurance).

There should not be so much glue that the glue runs out over the work and the bench and all over the craftsman. Anyway, a tight joint does not allow room for too much glue.

When gluing up a table or chair it is much better to glue up two opposite sections first and later glue them together. If everything is glued up at once, too many clamps are used, and it is more difficult to square the whole piece up at once.

Regardless of the variation of mortise and tenon joint you are making, or whether you are using power tools, the construction process is the same. Make the mortise first and transfer its dimensions to the tenon piece. But don't try to make the mortise and tenon independently. □

Hand-Cutting the Mortise and Tenon
Best results come directly from chisel and saw

by Ian J. Kirby

The mortise and tenon joint is used to bring two pieces of wood together, usually at a right angle, as in frames for carcases and doors, table legs and aprons, chair legs and rails. It is fundamental to woodworking and is made in innumerable variations, either by hand or machine. This discussion will focus on the basics of designing mortise and tenon joints to fit their purpose in a structure, and on making a single joint with hand tools. When there are only a few to do, a skilled workman will hand-cut them in the time it would take to set up machines.

A mortise and tenon joint gets its strength from the mechanical bond of letting one piece of wood into another, and from the adhesive applied to closely fitting long-grain surfaces. The craftsman must design the relative proportions of the mortise and tenon in order to best resist the forces the joint will encounter in service, to balance the wood tissue between the mortise and the tenon, and to maximize the long-grain gluing surfaces. Then he must make the parts accurately and cleanly, in order to achieve a close interface and thus a strong glue line.

A common mistake is to search in books for formulae and schematic diagrams of universal application. Instead, one should analyze the function of the joint in the structure one wants to make, and the loads it will have to bear. Does it have to resist downward pressure, or tensile load, or bending and twisting forces, or as is frequently the case, a combination of a number of forces? Knowing exactly what performance one wants from the joined pieces should be the first step in designing the joint.

The general rule is that there must be enough wood on the tenon, both in length and in section, to withstand the load it will have to bear. If the load is exclusively downward the tenon should be thick, but there is no need to have thick mortise cheeks. On the other hand, if there will be twisting forces, both the mortise cheeks and the tenon must be thick enough to withstand them. If the force is an outward pull or a pivoting, the tenon should be long enough to provide ample gluing area, or long enough to pass right through the mortise so it can be wedged on the other side.

The old rule of thumb when the mortise and tenon members are the same thickness is that the mortise should be one-third the width of the stock. This makes the tenon and each of the mortise cheeks the same thickness. Slavish adherence to this rule often leaves the tenons cut too mean, imbalancing the wood tissue between the tenon and the mortise cheeks. It would be better to say that the tenon thickness should about equal the thickness of the mortise cheeks added together. Thus in the example of a 1½-in. thick rail and stile, make the mortise cheeks each ⅜ in. thick and the tenon ¾ in., instead of making each ½ in. thick. The accompanying diagrams illustrate some of the forces such a joint commonly encounters, and some of the ways of keeping the joint strong where it needs to be strong. Notice that where twin tenons are used to resist bending and pivoting forces, as where the seat rail

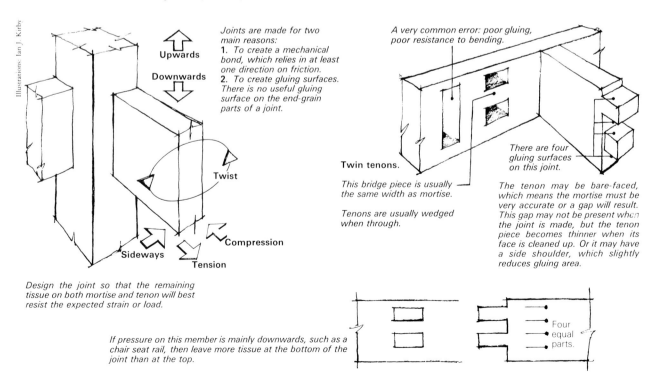

Illustrations: Ian J. Kirby

Upwards

Downwards

Joints are made for two main reasons:
1. To create a mechanical bond, which relies in at least one direction on friction.
2. To create gluing surfaces. There is no useful gluing surface on the end-grain parts of a joint.

Twist

Sideways

Compression

Tension

Design the joint so that the remaining tissue on both mortise and tenon will best resist the expected strain or load.

If pressure on this member is mainly downwards, such as a chair seat rail, then leave more tissue at the bottom of the joint than at the top.

A very common error: poor gluing, poor resistance to bending.

Twin tenons.

This bridge piece is usually the same width as mortise.

Tenons are usually wedged when through.

There are four gluing surfaces on this joint.

The tenon may be bare-faced, which means the mortise must be very accurate or a gap will result. This gap may not be present when the joint is made, but the tenon piece becomes thinner when its face is cleaned up. Or it may have a side shoulder, which slightly reduces gluing area.

Four equal parts.

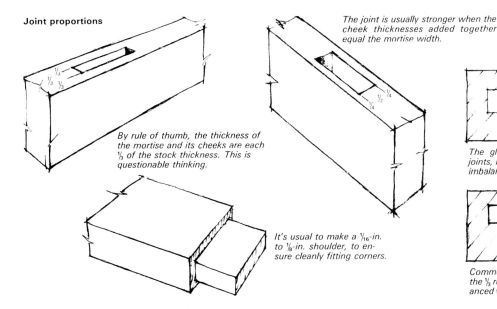

Joint proportions

By rule of thumb, the thickness of the mortise and its cheeks are each ⅓ of the stock thickness. This is questionable thinking.

It's usual to make a ⅟₁₆-in. to ⅛-in. shoulder, to ensure cleanly fitting corners.

The joint is usually stronger when the cheek thicknesses added together equal the mortise width.

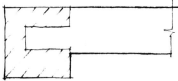

The gluing area is the same for both joints, but the bottom one is mechanically imbalanced.

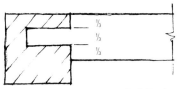

Common error: Automatically following the ⅓ rule leaves a weak tenon, and imbalanced wood tissue on the mortise cheeks.

meets the side rail of a chair, the object is to increase side-grain to side-grain gluing area. A common error when joining two pieces of wood this way is to make the tenon so that long-grain gluing surface is reduced and end-grain surface is increased—no help at all. Another consideration in maximizing gluing area is the depth that the tenon goes into the mortise member. If it is not to be a through tenon, one can safely mortise to within ¼ in. of the far side of the stock.

The simplest form of the joint is a *T*, where the mortise is somewhere in the middle of a length of wood—as in the lock rail of a door, or between the side rail and back leg of a chair. This article will focus on that situation, leaving the added complexities of joints at corners (which are usually haunched), joints in grooved or rabbeted pieces (usually with long and short shoulders), and joints that are wedged or pinned, for a subsequent discussion.

Tools for mortising — For accuracy, the mortise is usually cut first and the tenon cut to fit it. The essential tool is a mortise chisel, which determines the width of the mortise and therefore of the tenon. The mortise chisel differs from an ordinary bench chisel in that it is stoutly constructed to withstand heavy pounding with a mallet and levering, its blade is precisely rectangular in cross section, and there is no narrow waist where the blade meets the handle. The rectangular section of the blade makes the chisel somewhat self-jigging in action, so it will cut an accurate mortise. Its stout shoulder allows levering out of the waste. The ordinary bench chisel with beveled sides is most inadequate for mortising because it will twist, and may snap off at its narrow neck. Beyond this, there are various chisel patterns evolved by the branches of the trade, which amount to two main types: socket, where the handle fits into a socket in the blade, and tang, where an extension of the blade enters the handle. A tang chisel usually has a leather washer between the blade and the handle to cushion the recoil after the chisel is struck with a mallet. The socket also offers resilience and thus performs a similar function. The handle may be of a ring-porous hardwood such as ash, which is prone to splitting and therefore will be bound with metal ferrules top and bottom. Or, it may be a denser diffuse-porous wood such as box, and no ferrule is used at the

top. Or it may be a high-impact plastic, which is quite satisfactory.

The details of the handle and how it fastens to the blade are matters of personal preference. What does matter is that the blade be stout, truly aligned with the handle, and truly rectangular in section. All too often, even new chisels fail to fulfill these requirements, but they can usually be put right (see box, page 17).

The other necessary tool is a mortise gauge, and you cannot make the joint reliably accurate without one. It differs from an ordinary marking gauge in that there are two spurs, one of them movable. The distance between them is struck from the chisel itself and transferred to all the pieces of wood at the same setting. This critical distance can be maintained even when the position of the fence needs to be altered to account for mortise and tenon members of different thickness. A good mortise gauge is expensive, but it will last a lifetime if it is reserved for marking out mortise and tenon joints. To try to manage without one, by resorting to two settings of a marking gauge, is futile and plain bad practice. A mortise gauge often has a single spur on the side of the beam opposite the double spurs, apparently an encouragement to use it as an ordinary gauge as well. I usually remove or grind off this spur. In view of the expense of the gauge and its importance, it should not be expected to withstand the robust usage that a marking gauge is liable to receive.

Whether the fence is locked to the beam by means of a thumbscrew or a slotted screw is not important, but the life of the gauge will be considerably extended if this screw is not overtightened. It bears on the brass sliding strip that houses the moving spur, and there should be two small protective pellets of soft metal between the screw and the brass strip. If you have occasion to take the tool apart, be sure not to lose the pellets. If they aren't in there already, then make two and put them in. The spurs on the new gauge are usually ground to a cone-shaped point, as on a pencil. Although some workers like to sharpen them as if they were tiny knives, I believe the gauge is more accurate if they are left alone.

Cutting the joint depends on the direct relationship between these two accurate tools, the mortise chisel and the mortise gauge. No other tool need intervene between them in

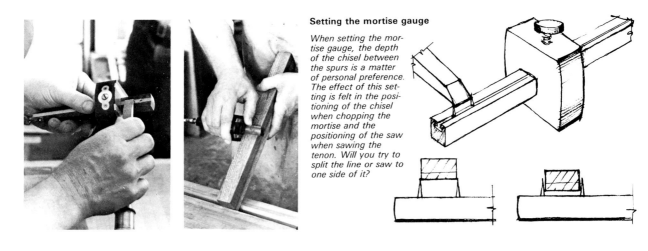

Setting the mortise gauge

When setting the mortise gauge, the depth of the chisel between the spurs is a matter of personal preference. The effect of this setting is felt in the positioning of the chisel when chopping the mortise and the positioning of the saw when sawing the tenon. Will you try to split the line or saw to one side of it?

quickly producing the most accurate joint. The mortise should not be widened by moving the chisel out of alignment at each cut, nor is it wise to adjust the width by paring its sides. Design consideration notwithstanding, one chops the mortise to the width of the chisel one has. The need for a set of several chisels quickly becomes apparent.

Setting the mortise gauge — There is a need for some fine judgment in deciding exactly how to set the mortise gauge from the chisel. It depends on how you intend to go about sawing the tenon: Will you try to saw to one side of the gauge line, or will you try to split it? To be in a position to be able to split the line, the spurs are set with the chisel between them, rather than with their very points exactly coincident with the chisel's extreme corners. This may seem like the workmanship of risk gone mad, but it does amount to the thickness of a line and can mean the difference between a good fit straight from the saw or one that needs further adjustment. The chisel should sit between the spurs about as deeply as the spurs will sink into the wood as they mark. This affects only the sawing of the tenon, not the width of the mortise, which is determined by the width of the chisel alone. The chisel will just touch the inside of each line and some of the gauge lines may remain visible after the cutting.

Having set the spurs, set the fence relative to the nearest spur to determine the cheeks of the mortise and the shoulder of the tenon, and mark them both on all the pieces of the wood. If the mortise member is thicker than the tenon piece, the fence setting will have to be changed, but on no account change the spur setting.

Shoulder lines — Shoulder lines are knifed round deeply with a try square from the face side and face edge. In the end, this knife line is the part of the shoulder that will show. It should be made with a thin knife sharpened flat on one side, like a chisel. I keep a small pocketknife for this job alone. The line should be crisp and deep, made with one pass of the knife. Shoulder lines are never made with a pencil since it leaves no register for subsequent paring with either chisel or shoulder plane. Scribing across the grain with a pointed tool is equally worthless, because it merely scratches the surface and drags up the wood fibers rather than cutting them.

Chopping the mortise — The mortising chisel, once it is correctly started, is self-jigging: each cut tends to follow the previous cut. However, care must be taken to chop vertically

or the mortise will wander. The important thing to get right is stance and body position relative to the workpiece and to the tool. The tool is held almost at arm's length and aligned with the center of the body. This way it is easy to see that it is vertical. It cannot be seen by standing over it. A good aid for the beginner is a straight piece of wood clamped to the face side of the work as an extension of its known accuracy, in advance of the joint itself so it doesn't get in the way. A less good aid is a try square resting on the bench against the work. This relies on the assumed flatness of the bench, rather than registry on the known accuracy of the workpiece. The square tends to fall over when the chisel is struck.

In any event, the workman must stand far enough back to sight the chisel properly, and to strike it hard with the mallet while continuing to sight it. The diagrams on the next page show the orientation of the chisel and the strategies for enlarging the mortise once it is begun. The most common fault is to strike too lightly. Cutting a mortise is quick, once one has enough confidence to strike each blow hard.

The best mallet for mortising is the cabinetmaker's or carpenter's type, which has a heavy rectangular head and a large, flat striking face. It has little tendency to deflect. One can confidently deliver a substantial blow and still keep one's eye on the cutting end and alignment of the chisel, not on its handle. Many people try to use the common cylindrical carver's mallet, which is meant for light tapping. Since the chisel handle is also domed, a good smash is likely to deflect onto the hand, also bruising the confidence.

Obviously, the workpiece has to be placed on the bench so that the correct stance can be taken, but its position is also important in other ways. The process involves heavy impacting with some risk to the bench surface, especially if the chisel accidentally cuts right through. The crucial part of the bench, for me anyway, is the surface in front of and around the vise, where the bench stop is. This should always be in perfect condition and truly flat, so it makes sense to do heavy pounding over the leg away from the vise. The workpiece can be clamped down, but with experience this becomes unnecessary. There is no need to support the cheeks of the mortise with clamps because the direction of the impact and of severing the fibers is such that (unless the grain is very wild) little strain will be put on the cheek tissue.

Because levering out the chips bruises the fibers at the end of the mortise, work it to full depth but to within only ⅛ in. of the ends. The ends may then be squared up to the line with one clean cut.

Align the chisel with the center of your body, strike it hard, and then lever out the waste.

Chopping the mortise

When the chisel is struck, it tends to cut into the wood tissue in the direction away from the sharpened side. A scooping action results, giving rise to two different methods of removing the waste from the mortise.

One method is a form of layering to achieve the required depth. The chisel is driven to the same depth in each position, with about ½ in. being the maximum depth, depending on the hardness of the wood. Move the chisel about ⅜ in. away from its flat side for each new cut. Repeat the process after the first bottom is made. Try to keep a level bottom without deep troughs, to avoid inadvertently chopping through.

Start

90°

Outline of proposed mortise

A common depth gauge is a piece of masking tape wrapped around the chisel.

Approx. center

The second method begins in the center of the mortise, turning the flat face of the chisel toward the center in each new position. The aim is to achieve the final depth with a wedge-like cut, and then to remove waste from top to bottom with each new position.

In either method leave the ends until last. Place the chisel in the knife cut you made when marking out the mortise, and use a small square to make sure the face of the blade is vertical. Drive the chisel accurately and hard; do not undercut the ends.

There are other methods of removing the waste, the most popular being with a drill the same diameter as the mortise width. This is to introduce yet another tool, which itself requires a setup and jigging to ensure exactness. Then the chisel has to be used anyway, whereupon the holes and the shape of the remaining waste encourage the chisel to twist. Others drill the waste and remove the residue by paring the cheeks with a wide bench chisel, invariably leaving the cheeks out of square or twisting or uneven. It is probably lack of confidence that persuades people that these other methods are safer and quicker when in fact they are neither. They invariably leave a worse result than that achieved straight from the mortise chisel. Resorting to such methods means only that the confidence that comes through practice is never acquired.

Sawing the tenon — Offer the tenon member up to the mortise to see how the gauge lines correspond to the actual hole, and to remind yourself of the decision you made when setting the marking gauge: Will you try to split the line, or to saw along one side of it?

Put the wood in the vise sloping away from you at an angle of about 60°. With the back (tenon) saw, begin the cut at the far end of the line across the top, that is, on the end grain. Watch the cut as it proceeds across the top to the near corner, and saw down the grain parallel to the end surface for about ⅛ in. This will create a good kerf in which the saw can be constantly registered as the cutting proceeds. Now saw down the long grain to the shoulder line on the face nearest you, without going any further down the back face but without lifting the saw out of the kerf at the back corner. This requires practice—the idea is to saw down only one line at a time, while keeping the saw correctly positioned at the start of the line to be cut next. Now turn the wood around in the vise and cut the other diagonal, keeping the saw teeth inside the kerf all the time. Finally, put the wood upright in the vise and saw straight down to the shoulder lines. The diagrams on the next

page should make the procedure clear. This method, once mastered, permits very fast and confident cutting. If you begin the cut at the top corner and proceed across the top and down one side at the same time, the saw is liable to wander, and corrective adjustment on one line usually puts the saw off on the other line.

To cut the shoulders, remove the wood from the vise and place it on a sawing board. Train yourself to saw about 1/16 in. away from the knifed line, in order to finish back to the line with a 1-in. paring chisel. The original knife line should be deep enough to locate the chisel as much by feel as by sight. Don't try to cut the full inch of chisel capacity, which with most woods takes too much pressure. Cut a half-inch of shoulder line, then move across half an inch. You'll find, of course, that the first cut will register the chisel for the next, a most helpful guide. The amount you can cut at one time depends on the species of wood, but the aim is to saw close enough to the line so that one chisel cut will finish the shoulder, yet not so close that the chisel can't easily click into the knifed line. If the tenon is wide, a shoulder plane is more practical than the chisel. But less than 4 in. of shoulder makes holding the plane somewhat more difficult.

Many people reason that a fine dovetail saw will produce a cleaner surface on the tenon. The Western-style dovetail backsaw, however, cuts on the push stroke and simply isn't stiff enough for the section of wood normally encountered in tenoning. The blade tends to buckle, inducing wander. The Eastern-style dovetail saw avoids this problem by cutting on the pull stroke, putting the blade into tension. However, it is

Sawing the tenon

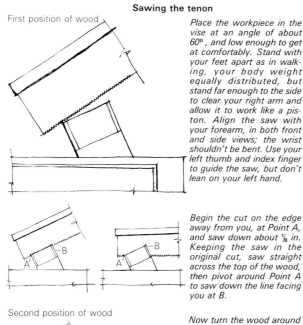

First position of wood

Place the workpiece in the vise at an angle of about 60°, and low enough to get at comfortably. Stand with your feet apart as in walking, your body weight equally distributed, but stand far enough to the side to clear your right arm and allow it to work like a piston. Align the saw with your forearm, in both front and side views; the wrist shouldn't be bent. Use your left thumb and index finger to guide the saw, but don't lean on your left hand.

Begin the cut on the edge away from you, at Point A, and saw down about ⅛ in. Keeping the saw in the original cut, saw straight across the wood, then pivot around Point A to saw down the line facing you at B.

Second position of wood

Now turn the wood around in the vise, at a similar height and angle to the first position. The area already cut is shaded. Place the saw in the kerf across the top of the piece, and saw down the line D. The saw pivots around Point C, and again must not be lifted out of the wood at C.

Third position of wood

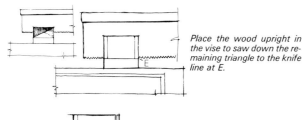

Place the wood upright in the vise to saw down the remaining triangle to the knife line at E.

Keep the work upright in the vise to saw down the side lines at F.

still a good deal slower than a tenon saw and has a distinct tendency to wander when sawing through a large section. The improvement in surface quality is marginal.

Other than the shoulders, the joint should not need trimming. The tenon should come directly from the saw and the mortise straight from the chisel. It is wrong to adopt the notion that on one hand it cannot be done, or on the other that one should leave a margin of safety by cutting everything oversize, to be trimmed right. The paring of a tenon, other than to make a minor adjustment, almost always puts it into twist, or removes too much from one side and thereby offsets the shoulders, or puts it out of alignment so it won't enter the mortise at 90°. It is far better to practice sawing and learn to saw correctly in the first place.

Checks — There are several ways to check the accuracy of the joint before it is put together. First verify that the faces of the tenon are in line by holding a rule against the side of the wood and sighting the tenon against it. Twist or angular misalignment will be apparent. For the mortise, first check the cut ends by placing a rule into it (or through it) so that it registers against the end-grain surface. The rule should touch the whole face at both ends—watch for a bump and make sure these surfaces have not been made concave by angling the chisel back. Next, make sure the ends are vertical by holding a try square up to the rule. Finally, check for twist in the cheeks by sighting into or through the mortise.

The joint should now be assembled and checked again, although a limited amount can be learned from a single practice joint. The real test is assembly of four joints into a rectangular frame, to which the following operational checks apply. First, hold a rule across the joint to see whether both mortise and tenon are in the correct plane. If they are not, subsequent gluing and cleaning up will be very difficult. See whether the shoulders pull up tight, that the shoulder lines are even and not offset, and that the whole assembly (or subassembly) is not in winding. Finally, see whether the two pieces (or all four in a frame) come together at a right angle.

Minor adjustments to correct any of these conditions can be made by careful paring with a wide bench chisel. There are pros and cons as to whether you adjust the mortise or the tenon, and it depends on the condition you are trying to put right, but in the main the tenon is easier to adjust. You can see more easily where the correct areas are from which to work, and where wood needs to be removed. The important

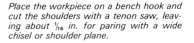

Place the workpiece on a bench hook and cut the shoulders with a tenon saw, leaving about ¹⁄₁₆ in. for paring with a wide chisel or shoulder plane.

Checking the tenon

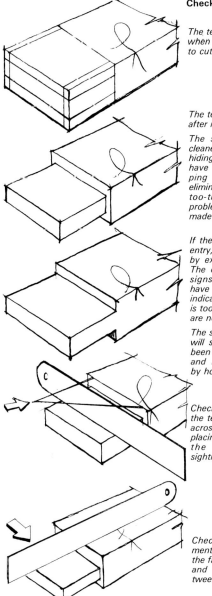

The tenon will look like this when it is marked out prior to cutting.

The tenon will look like this after it has been cut.

The shoulder will give a cleaner finish to the joint, hiding any tissue you may have bruised when chopping the mortise. It also eliminates the need for a too-tight fit in width, a problem when the joint is made bare-faced.

If the joint is very tight on entry, find the tight spots by examining its surfaces. The edge will often show signs of compression or have a glazed appearance, indicating that the mortise is too short or that its ends are not vertical.

The saw marks on the face will show how well it has been cut. Check with a rule and remove excess wood by horizontal paring.

Check the surface quality of the tenon by placing a rule across it. Check for twist by placing the rule parallel to the shoulder line and sighting over it.

Check accuracy of alignment by placing a rule on the face of the tenon wood and sighting the gap between it and the tenon.

Correcting new chisels

With an understanding of how the mortise chisel is used, it is easy to see that the tool's handle and blade ought to be in line, so it can be sighted vertically, and that the blade has to be exactly rectangular in section, so it can chop a square mortise. Many of these tools come from the factory out of line and out of square, inadequate for the task they are made to do. They can usually be put right, and it is crucially important to do so, but it may take several hours of corrective work.

If at all possible, buy mortising chisels in person, not by mail, and have an accurate try square with you. A 4-in. engineer's try square is most useful for this. Make sure that the back of the blade (opposite the sharpened bevel) is flat, then check that the handle is in line with the blade both in front view and side view.

An out-of-line chisel isn't useless, since you can compensate each time you sight up, but it is an added difficulty you could well do without. Repair it by removing the handle and fill the tang hole, then redrill it. This is not an easy task, and you may be better off making a whole new handle.

Now check the sides of the blade against the back. If the two sides and the back are not at right angles, the chisel will twist as it is driven, making the mortise wider than it should be, leaving a poor face on the cheek and inducing wander. No amount of compensating by gripping the handle tightly will stop this twisting. An out-of-square chisel is the result of sloppy manufacturing standards at the finishing stage. The only way to correct it is to grind the back face perfectly flat, and then to grind the two sides until they are at right angles to the back face. The front face is not as important, but it might as well be right as not since it will help in sharpening the edge square. A machine shop will be able to do the grinding for you, or you can do it on a coarse oilstone, or on a piece of carborundum cloth glued to the flat bed of a machine and lubricated with a little oil. Removing metal from the edges will make the chisel a little narrower than its nominal size. This is of little consequence. There is no good reason for the chisel to match any particular linear measurement, whereas it must be correct to angular measurement to perform.

---

thing is to analyze exactly where to remove fiber, and not to attack willy-nilly. The most controlled way to adjust a tenon is to put the work horizontally in the vise and pare horizontally across the grain. Do not pare in the direction of the grain, because the chisel will want to follow the long fibers and you are liable to remove far too much wood.

The crispness of the shoulder line is generally held to be the mark of success, but in a rectangular frame it is by no means the only thing. In particular, whether or not the frame actually is rectangular depends in part on the distance between shoulder lines. This makes adjustment of shoulders a very tricky process involving more than one joint that happens not to fit crisply. Check for squareness, not with a try square but by measuring the diagonals of the frame, which should be exactly the same.

In the glued-up frame, faults that arise from small inaccuracies within each joint manifest themselves dramatically as twist or wind or lack of flatness. For example, a tenon cut on the angle will result in a badly angled stile and probably a

twisted frame. The frame should be checked by sighting across from one member to another to ensure that they are parallel. If they are not, the correction, once the frame is glued together, requires planing the whole thing flat, a considerable task. Paying attention to the checks made on the individual joints can prevent such problems.

Clamping — The work is best clamped together on an already flat surface. Clamping blocks should be used to protect the wood and to direct the pressure to the shoulder lines. The more important interface, however, is the effective gluing surface between the sides of the tenon and the cheeks of the mortise, and it is usual to use a C-clamp and a pair of blocks to apply some light pressure here. All the places that cannot be reached by the plane after glue-up should be cleaned and polished before glue-up. □

Ian Kirby directs Kirby Studios, a school of woodworking and furniture design, in Cumming, Ga.

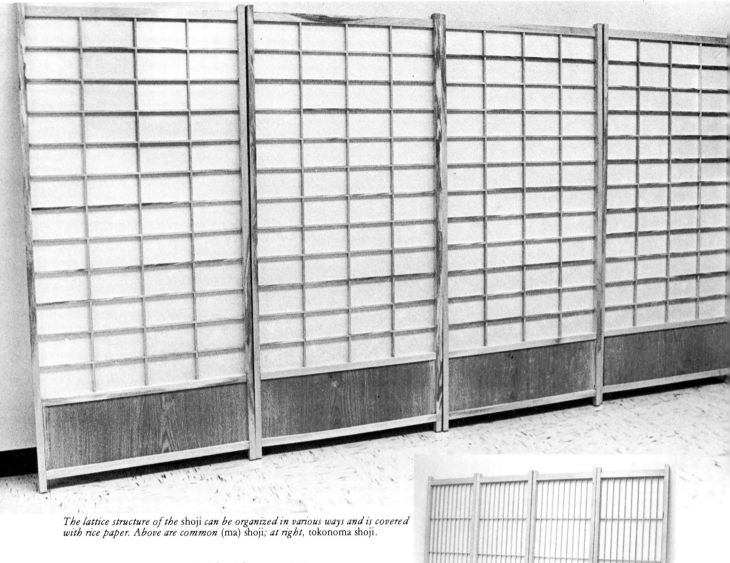

The lattice structure of the shoji *can be organized in various ways and is covered with rice paper. Above are common* (ma) *shoji; at right,* tokonoma shoji.

Japanese Sliding Doors
The traditional way to make *shoji*

by Toshio Odate

The traditional Japanese house allows for very flexible living. The house is post-and-beam, with the spaces between posts filled by doors, most of them sliding. Thus any wall, interior or exterior, can at different times become an entrance or an exit, a window or an open space, as the people desire. In this article I am going to show you the authentic way of making a sliding door. Although there are many kinds of sliding doors, the kind that Westerners associate with Japan are called *shoji*. This type of door consists of a softwood frame filled by a light latticework of thin wooden strips (called *kumiko*), to which is glued a layer of rice paper. The rice paper filters the light from outside into the home, for the people to enjoy.

When I was 16 years old, I was apprenticed to a *tategu-shi*, a maker of sliding doors. My master and I would carry the

tools on our shoulders from house to house and place to place. We often worked out-of-doors, under an overhang, or in a vacant cowshed. Everywhere we went we made the planing boards, beams and horses on which we could prepare the customer's materials, and when we were done, this equipment remained with the customer. We would stay at a single job for as short a time as one week, or as long as three months, working from dawn to dark, whatever the weather. After seven years of this, I could call myself *shokunin*, which means craftsman. Such an apprenticeship is the only way to acquire a skill in Japan, for these kinds of knowledge are nowhere written down and never pursued as hobbies. I don't imagine you can become *shokunin* simply by reading my article.

For that matter, I am no longer *shokunin*. I have been in America 24 years now, and my commitments are different.

But I am still a skillful person, and because of my unusual life I can be a bridge from the traditional Japanese way to the American craftsman who wants to understand. You may find new uses for *shoji* and other ways to make them. Each craftsman has his own experience and training. I can not tell you how to make American *shoji*, but I can describe for you how the *tategu-shi* has always made Japanese *shoji*. If you know where the design comes from, even if you change it to suit your own life, you will know what you are doing.

Varieties of sliding doors—*Shoji* is only one of the many kinds of sliding doors. The outermost door of a Japanese home is a wooden storm door (*amado*) which is closed tight every evening and left open during the day. Behind the storm door is a glass door in a wooden frame (*garasu-do*). The *shoji* is next. Often there is a narrow veranda, 3 ft. to 4 ft. wide, between the glass door and the *shoji*. This hallway borders the living space and is used to pass from one room to another (figure 1). Sliding room dividers separate the interior spaces. These room dividers can be *shoji* (with translucent rice paper), *fusuma* (with opaque paper and a very thin frame), *itado* (with wood panels), or a combination of *shoji* and *itado*. A living/dining room is commonly converted into a bedroom at night. Dining tables are folded flat, and beds are soft mattresses that are folded and stored every morning. Most rooms have a built-in closet with sliding doors (*fusuma* or *itado*) for household supplies.

The seven traditional styles of *shoji* are shown in figures 2 and 3. The one I will describe how to build is the common (*ma*) *shoji*, whose frame contains three vertical *kumiko* and either nine or eleven horizontal *kumiko*, with a hipboard (*koshi-ita*) at the base. This *koshi-ita* is a solid wood panel and is called "hipboard" perhaps because it is the height of your hip when you sit on the floor. The size of the hipboard varies according to the total height of the *shoji*, but the spacing between the horizontal *kumiko* depends on the two sizes of rice paper available: 28 cm wide and 25 cm wide. The edges of the paper overlap on top of the horizontal *kumiko*. The wider paper is used with the nine horizontal *kumiko* to produce the classic *mino* proportions. The narrower paper is used with the eleven horizontal *kumiko* to produce the more contemporary *hanshi* proportions. These two variations, and more, are possible in all the styles of *shoji*.

Rice paper (*shoji gami*) sometimes is watermarked with a pattern, commonly of plum trees, blossoms, pine trees, bamboo leaves or chrysanthemums. Sometimes these patterns are realistic, sometimes abstract. Because they are watermarks, you can see these patterns best from the inside of the room when daylight passes through the paper. The effect is like sitting with a beautiful garden outside, the pattern on the paper like the shadows of trees and flowers. Bringing nature inside the home is characteristic of the Japanese. The cultivation of miniature trees, *bonsai*, is another example of this.

Preparation—It is a common saying among Japanese craftsmen that when an apprentice can accurately prepare door materials he knows how to make a simple sliding door. People see the finished product and they say, "He is neat," or "He has skill," but actually most of the quality of the work is in the preparation of the materials. For typical dimensions of the parts, refer to figure 3. I begin with the hipboard. If you do not have stock wide enough to make it out of one piece,

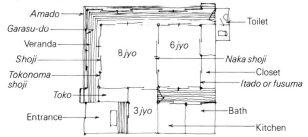

Amado • Garasu-do • Veranda • Shoji • Tokonoma shoji • Toko • Entrance • 8 jyo • 6 jyo • 3 jyo • Toilet • Naka shoji • Closet • Itado or fusuma • Bath • Kitchen

Sliding doors allow for a flexible living area. Jyo is 6 ft. by 3 ft., the size of the grass mats (tatami) *used to cover the floor.*

Fig 2: Varieties of *shoji*

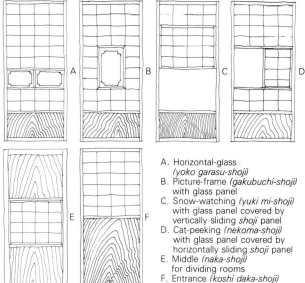

A. Horizontal-glass (*yoko garasu-shoji*)
B. Picture-frame (*gakubuchi-shoji*) with glass panel
C. Snow-watching (*yuki mi-shoji*) with glass panel covered by vertically sliding *shoji* panel
D. Cat-peeking (*nekoma-shoji*) with glass panel covered by horizontally sliding *shoji* panel
E. Middle (*naka-shoji*) for dividing rooms
F. Entrance (*koshi daka-shoji*) currently out of fashion

Fig. 3: A typical *ma-shoji*
There are two alternate top rail designs. The rail can be 1⅝₁₆ in. thick and rabbeted to fit the track it slides in, or ¾ in. thick, unrabbeted. The thicker rail looks more finished, because the rabbet covers the track. The thinner rail is stronger, because the tenon can be wider.

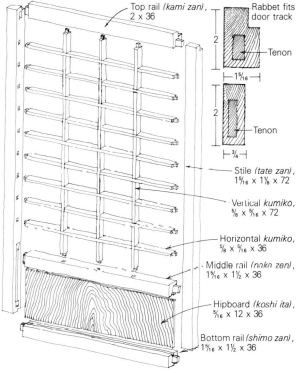

Top rail (*kami zan*), 2 x 36
Rabbet fits door track
Tenon
Tenon
Stile (*tate zan*), 1⁵⁄₁₆ x 1⅛ x 72
Vertical *kumiko*, ⅝ x ⁵⁄₁₆ x 72
Horizontal *kumiko*, ⅝ x ⁵⁄₁₆ x 36
Middle rail (*naka zan*), 1⁵⁄₁₆ x 1½ x 36
Hipboard (*koshi ita*), ⁵⁄₁₆ x 12 x 36
Bottom rail (*shimo zan*), 1⁵⁄₁₆ x 1½ x 36

The tategu-shi (sliding-door maker) begins by planing his stock to size, left. He uses planes that cut on the pull stroke, and he supports the wood on a kezuri-dai, that is, a beam held at one end by a triangular support and lodged against anything sturdy at the other. A nail driven into the beam stops the work against the pull of the plane. Traditionally, the kezuri-dai is fashioned at the work site and left behind when the craftsman finishes the job and moves on. Layout, above, is done with a thin, narrow square and a marking knife. Similar pieces, here both stiles for one shoji, are clamped together and layout lines struck across the stack.

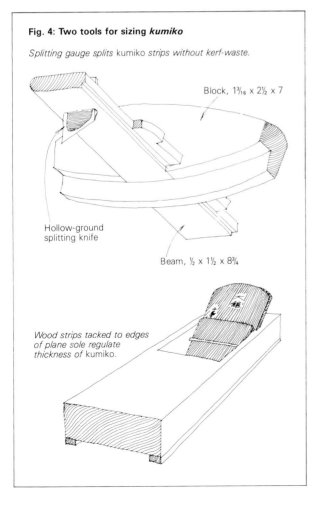

Fig. 4: Two tools for sizing *kumiko*

Splitting gauge splits kumiko *strips without kerf-waste.*

Block, 1³⁄₁₆ × 2½ × 7

Hollow-ground splitting knife

Beam, ½ × 1½ × 8¾

Wood strips tacked to edges of plane sole regulate thickness of kumiko.

you begin by gluing it up. This way the glue will be dry when it comes time to plane the hipboard and cut it to size.

Next I prepare the stiles. The front face, which will face out of the room and receive the paper, must be planed flat and free of twist. Next plane the inside edge perpendicular to the front face, but instead of being straight along its length, it should bow slightly. This will hold the stile tight against the *kumiko*. The large tenons of the rails will be made to fit tight in the mortises of the stile—so tight that they will have to be hammered home. But the *kumiko* are delicate. Bowing the stile to press gently against the *kumiko* shoulders, instead of making the tiny mortises and tenons hold the parts snug, I call the "thoughtfulness of the craftsman."

Once the front face and inside edge are planed, gauge the width of the face with a marking gauge and plane the outside edge. Then gauge the thickness of the edge and plane the back face. This is the face that will show in the room. All the frame parts of the *shoji* are planed in this order. Plane the *kumiko*-facing edges of the top and middle rails and the inside edge of the bottom rail to bow in. Now I cut the stock to rough length and turn to the *kumiko*.

Apprentices being trained today use a tablesaw and a thickness planer for preparing *kumiko*. The hand method I describe here is the one I learned. I begin by planing perfectly flat a 1-in. thick redwood board, 6 ft. long. Mark it to ⅝-in. thickness, and plane the back face to the mark. Plane the edges of the board square to the face.

Now I use a splitting gauge (figure 4), which is like a marking gauge but larger and heavier. I score the board, first one face, then the other, until I can snap off the *kumiko*. I plane the edge of the board again, then split off another *kumiko*, repeating the process until I have plenty of extra pieces.

Next I wet the knifed surface of each piece with a damp

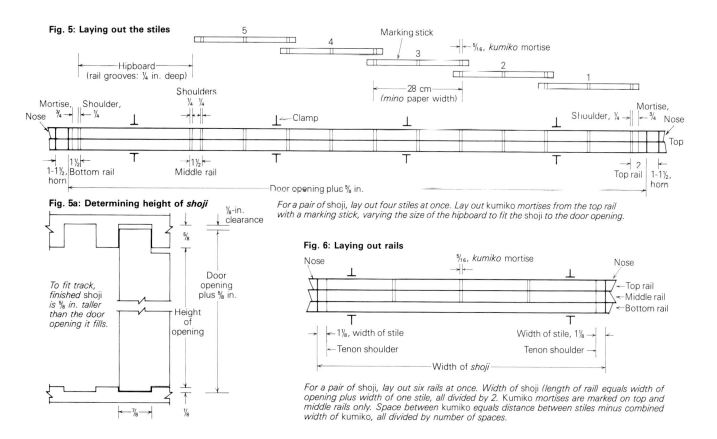

Fig. 5: Laying out the stiles

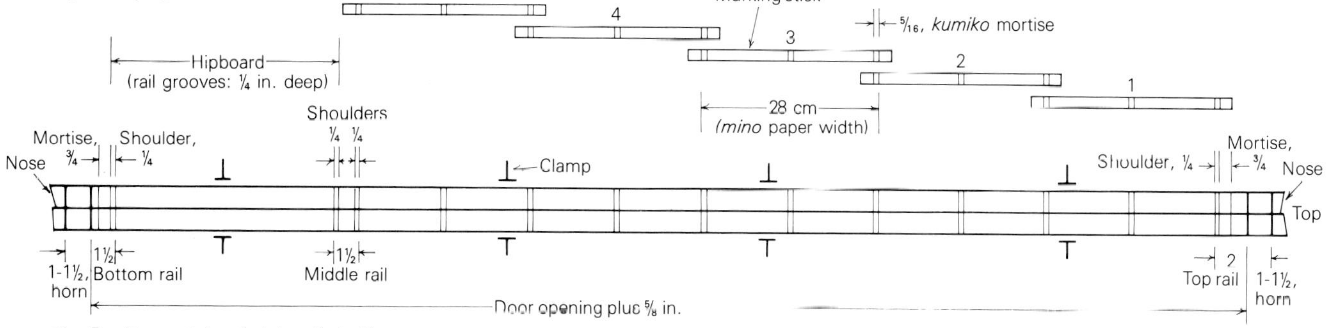
5
4
3
2
1
Marking stick
5/16, kumiko mortise
Hipboard
(rail grooves: 1/4 in. deep)
28 cm
(mino paper width)
Shoulders
1/4 1/4
Mortise, 3/4
Shoulder, 1/4
Nose
Clamp
Shoulder, 1/4
Mortise, 3/4
Nose
Top
1-1½, horn
1½
Bottom rail
1½
Middle rail
2
Top rail
1-1½, horn
Door opening plus 5/8 in.

For a pair of shoji, *lay out four stiles at once. Lay out* kumiko *mortises from the top rail with a marking stick, varying the size of the hipboard to fit the* shoji *to the door opening.*

Fig. 5a: Determining height of *shoji*

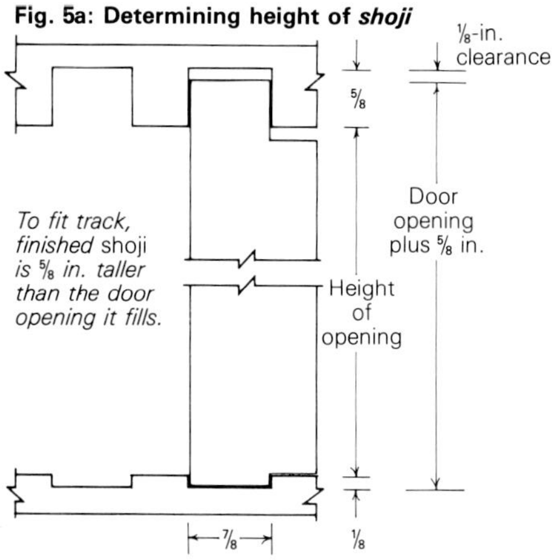
1/8-in. clearance
5/8
Door opening plus 5/8 in.
To fit track, finished shoji is 5/8 in. taller than the door opening it fills.
Height of opening
7/8
1/8

Fig. 6: Laying out rails

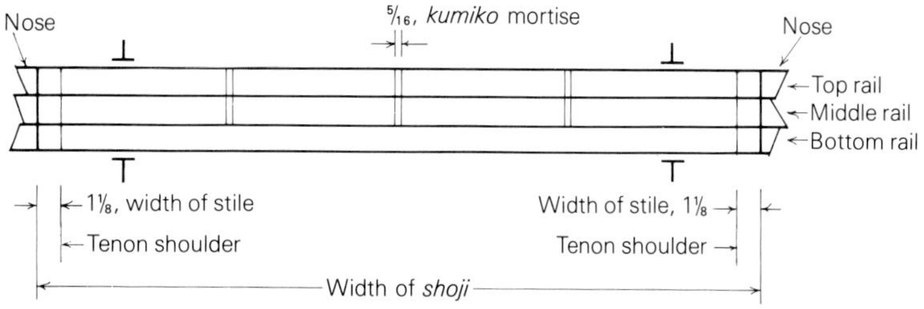
Nose
5/16, kumiko mortise
Nose
Top rail
Middle rail
Bottom rail
1⅛, width of stile
Tenon shoulder
Width of stile, 1⅛
Tenon shoulder
Width of shoji

For a pair of shoji, *lay out six rails at once. Width of* shoji *(length of rail) equals width of opening plus width of one stile, all divided by 2.* Kumiko *mortises are marked on top and middle rails only. Space between* kumiko *equals distance between stiles minus combined width of* kumiko, *all divided by number of spaces.*

cloth, to relieve the pressure made by the knife. If you don't do this, the *kumiko* will eventually swell after they are assembled, and cause trouble. I lean the *kumiko* against a wall so the air can move around them until they are dry, and then plane the split edges square. To make sure they will be exactly the same width, I plane three or four *kumiko* at once, using a plane I reserve for this purpose. It has wood strips tacked to the bottom to stop the cut (figure 4).

Laying out the joints—The wall opening and tracks built by the house carpenter determine the outer dimensions of the *shoji*. The width of the rice paper determines the spacing of the horizontal *kumiko*. Marking out this spacing from the top rail determines where the middle rail goes, and thus the *tategu-shi* finds the height of the hipboard. All other measurements are according to the discretion of the craftsman. The measurements in the drawings are typical.

For speed and accuracy you lay out similar pieces, both stiles for instance (or four, when one opening requires two *shoji*), at the same time. Use clamps to keep the pieces aligned. The *tategu-shi* uses his clamps mostly for layout, almost never for assembly. I strike finished mortise heights and tenon shoulders across the width of the stock, using a square and a marking knife. Pencils and pens are not so accurate as the knife, and are used only for marking to rough length. I mark the stiles first, then the rails, then the *kumiko*.

The stiles: It is customary to orient the stiles the way the wood grew in the tree. So, I make sure the largest growth rings are at the bottom of the stiles when I start to lay them out. Clamp the stiles together, inside edge up, and mark the finished height of the *shoji*, ⅝ in. longer than the height of the opening it will fill. The extra length fits the tracks, top and bottom, in which the *shoji* will slide (figure 5). Next I make a mark for the horns, 1-in. to 1½-in. past the finished height on either end. Most of the horns will be cut off later, but for now they keep the stiles from splitting when the rail tenons are driven into their mortises, and they also protect the ends of the stiles from damage during the work. Mark the width of the top and bottom rails next, and within those widths mark the mortise height.

Next I mark off the mortises for the *kumiko*, using a marking stick. The stick carries the width of the paper and the position of three *kumiko* in relation to that width. Figure 5 shows the layout of *kumiko* mortises for the *mino*-size paper, 28 cm wide, which gives nine horizontal *kumiko*. The stick has two *kumiko* mortises marked just inside the paper width, plus one centered between them. I begin at the mark for the top rail, overlap it the width of one *kumiko* mortise, and knife off the other two mortises. Reposition the stick to overlap the last mortise marked, and mark the next two mortises. I continue in this manner five times, until I have marked off nine *kumiko* mortises, and then I mark off the top of the middle rail. Finally I mark the width of the middle rail, and within it the mortise height. I square all these knife marks across all four stiles, saw off the noses (the waste beyond the horns), unclamp the stiles and chamfer the ends against damage.

The rails: When two *shoji* fill a door opening, they overlap each other by the width of a stile. The width of each *shoji* thus is figured by adding the width of a stile to the width of the opening and dividing by two. The final rail length will be shy of this dimension, because the tenon is not quite a through tenon, but for now, I clamp the rails together, inside edge up, and mark their length as the *shoji* width (figure 6). Next I mark off the width of the stiles, which locates the tenon shoulders. The mortises for the vertical *kumiko* are marked next, equally spaced between the two stiles. I use a

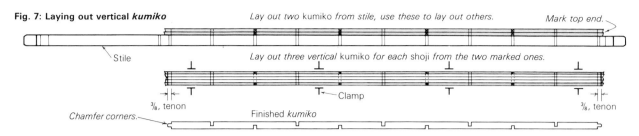

Lay out two kumiko from stile, use these to lay out others.

Mark top end.

Stile

Lay out three vertical kumiko *for each shoji from the two marked ones.*

Clamp

³⁄₈, tenon

³⁄₈, tenon

Chamfer corners.

Finished *kumiko*

Odate clamps the kumiko *(the thin strips that form the* shoji *grid) together in a stack and saws the notch shoulders a hair more than halfway through. A piece of scrap starts the saw correctly.*

With the kumiko *still clamped together, Odate makes their tenons. First he saws the shoulders, then breaks off the waste with a chisel pushed in from the end grain (top). The index finger on the bottom of the chisel acts as a stop to protect the shoulder. He cleans up the tenons by paring with the chisel across the grain (above). When all the tenons have been formed and their top and bottom corners chamfered with a few strokes of a plane, he unclamps the stack, fans out the* kumiko *and chamfers the other two corners (below).*

To break out the waste Odate pulls the corner of a flat chisel along the kerf (top). Then he clears the waste with a mortise chisel run in the notch, bevel-side down (above).

marking stick, as for the horizontal *kumiko*. Lastly I saw off the noses squarely, and unclamp the rails.

Vertical kumiko. To lay out the tenons and the notches for the half-lap joints on the vertical *kumiko* I transfer the layout lines from one of the stiles to two of the *kumiko*, and then from these two to the rest of the *kumiko* (figure 7). I clamp the two marked *kumiko* on either side of the stack of un-marked *kumiko* to strike the layout lines across the stack. It's a good idea to make two extra *kumiko* and not use the marked *kumiko* in the finished *shoji*. The vertical *kumiko* get notched alternately front and back. So I square every other notch around the underside of the stack. Last, I mark the tenon shoulders and lengths.

Now, while the vertical *kumiko* are still clamped together, I saw the notches and the tenons (photos, facing page). You cut both shoulders of one notch first, using a piece of scrap to start the saw correctly. Saw a hair more than halfway through the *kumiko*, break out the waste with a chisel, and clean up by running a mortise chisel, bevel-down, along the bottom of the notch. I insert a scrap of *kumiko* in this notch as security in case a clamp shifts while I'm cutting the other notches.

Next I cut the clamped *kumiko* stack to final length. To cut the tenons, square the shoulder lines around all four sides of the stack. Gauge the tenon on the end grain of the stack and on the faces of the two outside *kumiko*, then saw the shoulders. These are small tenons, so instead of sawing in from the end grain to meet the shoulder, I use a chisel to break off the waste. My index finger on the underside of the chisel acts as a stop to keep the chisel from damaging the *kumiko* shoulders. In all but the straightest-grained stock, I break a little bit wide and pare the tenons to the line.

Before removing the clamps, I chamfer the upper and

Fig. 8: Laying out horizontal *kumiko*

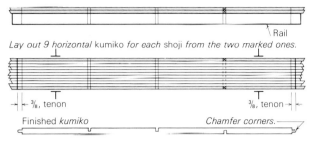

Lay out two kumiko *from rail, use these to lay out others.*

Rail

Lay out 9 horizontal kumiko *for each shoji from the two marked ones.*

← ³⁄₈, tenon   ³⁄₈, tenon →
Finished *kumiko*   *Chamfer corners.*

lower edges of the tenons. Then I remove the clamps and fan out the stack to chamfer one corner, then the other. The vertical *kumiko* are now ready.

Horizontal kumiko: Many people think that the *kumiko* overlap, every other one, as if they are woven. But *kumiko* will not bend that much. They are only partly woven. When there are three vertical *kumiko*, for instance, the notches in the horizontal *kumiko* are two, adjacent, on one face, one on the other (figure 8). They are marked out and cut exactly like the vertical *kumiko*.

Cutting the joints—The *kumiko* notches and tenons have been cut while the *kumiko* were clamped up for layout. The joints on the stiles and rails are now cut on the pieces individually, mortises first, then tenons. Cutting the tenons last lessens the danger of damaging them. The quality of a craftsman's skill is judged by his speed and accuracy. It is considered most important to make each cut with the saw or chisel the final cut—you go directly to the layout line. The

Like the planes, Japanese saws cut on the pull stroke. The long handle is usually held with two hands, spaced well apart for maximum power and control. There are three basic sawing stances, each suited for a different sort of cut. For crosscutting, left, Odate supports the stock on two low horses, holds it steady with his foot, and saws through. For sawing shoulders, as at top left of facing page, Odate sits so he can see where

the cut has to stop. And for ripping, as for tenon cheeks, center and right, he supports the stock on one horse so he can see the layout lines on the near edge and on the end grain at the same time. To avoid cutting into the shoulder, Odate saws on an angle into the near edge first, then turns the stock over to cut into the opposite edge, finishing with the saw straight up and down.

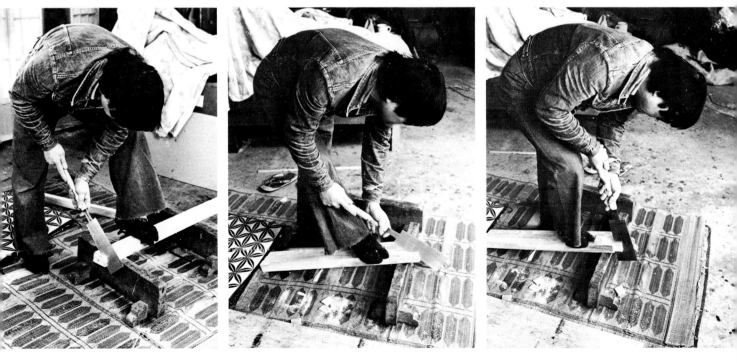

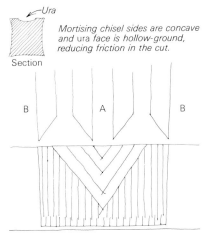

Ura

Mortising chisel sides are concave and ura face is hollow-ground, reducing friction in the cut.

Section

B   A   B

A: Begin in the middle of the mortise and chop out toward the ends, alternating sides, always with the ura face toward the middle.
B: At ends of the mortise, turn ura face around.

The tategu-shi *sits on the wood to steady it while he mortises it, above, stabbing his chisel frequently in a box of cotton wadding soaked with vegetable oil, to reduce friction. The chisel has three concave sides and a hollow-ground face (figure 9 and photo, top of facing page). He chops from the middle out, always with the face toward the middle of the mortise, except for the final cuts at either end of the mortise. These are angled slightly from the perpendicular (photo, right, includes a square for illustration only) to taper the mortise for a tight fit when the tenon is driven in.*

least contact lessens the chance of error and keeps the work crisp. Should the *shokunin* make a mistake, no matter how small, his error remains in the work, and even if only he knows it, it is a permanent reproach. Nothing can be done about it. So you learn not to make mistakes.

Japanese mortises are somewhat different from Western mortises, and so are some of the tools used to cut them. To get maximum strength in a delicate frame, the *tategu-shi* shapes his mortises with walls that taper in, to compress the fibers of the tenon without crushing them. The natural springiness of the wood enhances the mechanical strength of the joint. He works to very close tolerances: a shaving here or there makes all the difference. It is thought coarse to show end grain, so through-joints are used only in heavy entrance doors, rain doors, and doors that carry glass. For strength and refinement, the main joints of the *shoji* must be as deep as possible without going through. The bottom is paper-thin, thin enough for light to show through. But no mark must show on the outside. One slip and the wood is ruined.

You gauge the mortise width, making sure the fence is always on the front face of the stock, then chop your mortises with a chisel exactly as wide as the mortise. Japanese mortise

chisels are rectangular in section and will not turn in the mortise. Three sides are slightly concave, and the face, called *ura*, is hollow-ground (figure 9 and photo, facing page). This reduces friction in the heavy cuts. Stabbing the chisel frequently into a box of cotton wadding soaked in vegetable oil further reduces friction. The edges of the chisel scrape and true the long-grain sides of the mortise.

The *tategu-shi* strikes his chisels with an iron hammer, not the wooden mallet used in the West. He works from the middle of the mortise out, alternating cuts at either end, the *ura* always facing toward the middle. As the chisel cuts, it follows the bevel, so each cut shears toward the middle. As you near the ends of the mortise, you turn the *ura* around and chop straight down. The last cut at each end is with the chisel tilted slightly into the mortise. This tapers the walls just enough to pinch the tenon when it is driven in.

The *tategu-shi* does not lever waste out with the chisel as he chops, as does a Western woodworker. Instead he uses a small harpoon-shaped tool called *mori-nomi* (photos, facing page). Its face is flushed against the mortise wall, and the tool is tapped down and quickly jerked up. Its hook catches the chips and clears them out. Chopping, alternately with the mortise chisel and the *mori-nomi*, proceeds quickly until the final depth is approached. Then you slow down and gauge the depth with a piece of *kumiko* cut to length. Score the remaining wood with the chisel, and remove the last fibers from the bottom of the mortise with a *sokozarai-nomi*, another tool I have not seen in the West. It is a thin, goose-necked tool with a small spade-like bend at its end. This tool is not tapped with a hammer, but used like a scraper, with one or two hands, to level the bottom of the mortise.

The *kumiko* mortises need not be as deep as the mortises for the rail tenon because the bow in the stiles and rails holds them tight. The *kumiko* mortises are too small to be scraped in the usual way. So, you chisel to within ⅛ in. of the final depth and use a small steel rod to tap the wood down for clearance (about ¹⁄₁₆ in. deeper than the length of the *kumiko* tenon). This method works best in softwood.

The tenons on the rails are cut in much the same way as they would be in the West, although the *tategu-shi* holds his work differently and uses Japanese saws which cut on the pull stroke. First extend the shoulder lines (marked on the inside edge when the stiles were clamped together for layout) around the other three sides of each rail. Gauge the tenon thickness on the two edges and on the end grain. Saw the shoulders first, on all the rails, then line the rails up to saw the cheeks. The photos on p. 55 show how to proceed, sawing with the stock supported at an angle, so you can see the lines on both the end grain and the edge of the stock. Saw at an angle to the near edge of the shoulder, then turn the stock over to finish. This way there is less danger of oversawing into the shoulder. To cut the narrow third and fourth shoulders, you should not saw the shoulder right on the line, because the set of the saw can damage the first two shoulders sawn. Instead, saw a little wide of the shoulder and trim with a chisel. All shoulders cut, saw the length of the tenon ⅛ in. to ¹⁄₁₆ in. less than the depth of the mortise. Finally, chamfer the end of the tenon so it will go in easily.

Last, plow-plane the grooves that will hold the hipboard, and rabbet the bottom rail (the top too, if you are using the thicker top rail), so the *shoji* can fit into its track in the wall opening (figures 3, page 19, and 5a, page 21).

Assembly—The Japanese prefer natural surfaces. The *shoji* receives no finish except a final planing of all its parts, to clean them from handling. The finish plane takes off the slightest shaving, with only one or two passes. Pressing the plane hard against the stock burnishes the surface and brings the wood to a warm glow.

Cut and plane the hipboard to fit, allowing room for the wood to move across the grain, finish-plane its two faces, and chamfer all its edges. Finish-plane all other exposed surfaces of all other parts and lightly chamfer the edges of the main frame parts, except the inside front edges, where a chamfer would create a gap between them and the *kumiko*. Now at last, you're ready to assemble. I use rice glue that I make myself, so the *shoji* can be taken apart if it ever needs repair. Any starch glue, like wallpaper paste, will do.

Assemble the *kumiko* first. Group the horizontals together and the verticals together to make quick work of applying the rice glue to the shoulders of all the notches. Do not put any glue in the bottom of the notches, because glue here would prevent the *kumiko* from fitting tightly. Tap the *kumiko* together using a hammer. Fit the assembly into the mortises

The tategu-shi's *mortising tools. From left to right, a mortising chisel, with hollow-ground face (ura); a* mori-nomi, *whose harpoon-like hook is tapped down and jerked up to remove chips; and a* sokozarai-nomi, *which scrapes the mortise bottom flat and also lifts out chips. Last is a steel rod for tapping flat the bottoms of small mortises.*

Removing chips with the mori-nomi.

Scraping the bottom of the mortises with the sokozarai-nomi.

The kumiko *lap joints alternate, above, and must be eased into place. Below, Odate holds a small* shoji *he made for demonstration during a weekend workshop.*

in the top rail. No glue is needed here. Fit the hipboard into the groove of the middle and bottom rail, and fit the *kumiko* assembly into the mortises in the middle rail.

Now you are ready to add the stiles. First take a hammer and tap around the mortises so the edges of the rail shoulders will fit tight. Then apply glue to both stiles at once. Tap the rail tenons into one of the stiles, stopping when the *kumiko* tenons just begin to engage, then start the other stile in the same way. Make sure both stiles are going on straight, and tap them home with a hammer. Hammer on a small block of wood with chamfered edges to avoid damaging the stock. When the tenons fit tight, check to be sure the *shoji* is square and flat. Tap and twist it into shape if it is not.

Installation—With assembly, the tense part of the *shokunin's* challenge is accomplished. Installation is the joy of displaying your work. Place the *shoji* on the outside ledge of the bottom track and check the stiles against the door frame for alignment. Cut the bottom horns as close to the bottom rails as possible, but if necessary at slightly different heights to align the stiles parallel to the door frame. Rabbet the horns, like the bottom rail, to fit the groove in the track. Now put the *shoji* back on the ledge (not in the groove yet) and press the top of the *shoji* up against the outside of the top track. Make a mark on the inside face of each horn where the track meets it, add ⅝ in. to this mark, and you will have the length to which the top horns should be cut. Once they are cut, rabbet them to fit the track.

Applying paper—Rice paper has a smooth side, which is on the inside of the roll, and this should face out when the paper is applied to the screen. The horizontal strips are pasted to the *shoji* with rice glue, and they overlap one another like shingles, so the seams will not collect dust. Paper is traditionally applied by the housewife and customarily changed during the last week of the year, so that the paper is bright white for New Year's Day, signifying a fresh start. Old paper is easily removed by moistening it.

Besides the traditional *mino* and *hanshi*-size rolls, paper companies now make rolls one meter wide to be pasted on vertically in one piece. This opens up many possibilities in the spacing and patterns of the *kumiko*, which have always been carefully positioned to accommodate the traditional-size papers. This kind of change creates freedom in design, but it raises questions about pride in craftsmanship.

*   *   *

Well, finally you have finished and neatly installed a pair of *shoji*. You can appreciate now their character. The *shoji* paper draws in not only light, but light's warmness, softness and taste. The frames and *kumiko* that support the paper are not heavy or coarse. You open and close the *shoji* gently. The *shoji* has everything you need to feel peaceful. You retreat from the bustling world outside, you take off your shoes when you enter your home, and you sit down on a thin mattress in a room of *shoji* walls. You can call this place an oasis of life. □

Toshio Odate was trained as a tategu-shi *(maker of sliding doors) in Japan. Today he is a sculptor and lives in Woodbury, Ct. Odate's book,* Japanese Woodworking Tools: Their Spirit, Tradition and Use, *was published in 1984 by The Taunton Press. Drawings by the author.*

Locking the Joint

Tenons tusked, draw-pegged or wedged will hold without glue

by Ian J. Kirby

This article is about three variations on the mortise-and-tenon joint: draw-pegging, tusk tenoning and wedging. All exhibit a lively and creative mind on the part of their originators. Each has been in use for a long time and examples may be found in building construction, using material of very large section, as well as in the most refined cabinetmaking. The important lesson is that once the nature of the material is understood, and the principles of woodworking are applied, all the rest is wide open to the imagination of the maker. Superb execution of the work plays a supportive but secondary role in achieving a fine result.

The notion of a square peg in a square hole—the mortise-and-tenon joint—has been with us for thousands of years. Many woodworkers seem inhibited by this history and feel there must be a set procedure, developed over centuries, for any given joint. But joint design should not be looked upon in this light. There are two separate considerations.

One is, exactly what form shall a joint take? The maker must analyze the work the joint has to accomplish. He must consider gluing surfaces and strength of tissue on each member being joined. The second consideration, which usually gets seriously entangled with the first, is, does the workman have the skills and tools to make the joints he designs? The available tools at times can have a most consequential effect on the final outcome. Between these two considerations lies one of the more serious issues in teaching and learning to design. Our usual reaction as woodworkers, when we have to design anything, is to discard all those things lying outside our manufacturing ability. This tendency creates a form of tunnel vision. On the other hand, the serious design student without much woodworking skill is unfettered by tunnel vision. If he understands the principles, a creative solution is frequently the outcome. Conversely, the woodworker who has the skill to do the work but who ignores the principles is quite liable to make the thing in an accurate manner, while making it fundamentally bad. Fortunately, neither extreme is usually the case—we are mostly somewhere in between.

The idea that superb craftsmanship is the necessary springboard for all other areas of knowledge can thus be seen to be questionable. Indeed, the reverse is more likely to be the case—I would encourage any woodworker, beginning or advanced, not to feel inhibited by a lack of specific information about joints and other woodworking details. Knowledge of the principles involved is the salient factor—specific detail can usually be found in books and magazines and applied as the need arises. When a ready answer is not available, an intelligent application of the principles will normally solve the problem. We have access to a great deal of knowledge by studying the work of our predecessors, but that is not to say there is no room left now for the design of new joints and new systems by the individual woodworker. It is simply a question of the combination of basic principles with creative thinking,

brought to fruition by excellent workmanship.

Most joinery in solid wood offers the choice of hiding the structural nature of the work, or of putting it on view. We have through dovetails and secret mitered dovetails, through tenons and blind tenons. Exposed structural elements may be embellished, usually by carving in various ways or by the introduction of alternate woods for wedges and pins. This type of joint may have such an effect on the visual quality of a piece of work that its whole nature is dramatically changed. A similar effect can be seen on a large scale in architecture, in such things as hammer-beam roofs and exposed girder work, and on a small scale in furniture joints. The drawings on the following two pages show the basic forms of these furniture joints and the basic procedure for making them.

These joints have the unusual characteristic of being able to function well without glue. Two of them—the tusk tenon and the draw-pegged tenon—can be disassembled without too much trouble. All three, once made, can be put together and worked on immediately—they are mechanically locked. The tusk tenon and the draw-peg tenon need no clamps at all, and the wedged mortise and tenon can be clamped until the wedges are driven, and then the pressure can be removed.

Good examples of draw-pegging can be found in old-style post-and-beam houses, and the joint is of course used by enthusiastic builders today. It is traditionally made by offsetting the drilled holes so the entry of the peg draws the structure tightly together. But I don't think it is cheating too much in cabinet work to put the work into clamps, apply normal gluing pressure, and drill through both parts all at once. For that matter it can be done at the time of gluing (unless of course you want the joint to knock down). If you draw the joint together in a small cabinet, don't overdo the off-centering. The result might be an unsightly mess on the emerging side of the peg hole, and excessive strain along the shear lines of the tenon.

In the wedged joint, be accurate when setting out and sawing the wedges. If they are too small the joint will be loose, and if too large the wood fibers of the tenon will be crushed and broken. Don't make the tenon any more than $1/16$ in. longer than the depth of the mortise. It is a total energy waste when it comes to cleaning up the joint—unless you intend to carve and feature the protruding tenon.

When tusk-tenoning, the critical points are an accurate match between the slope of the wedge and the hole through which it passes, and enough space inside the mortise for the wedge to do its work. This joint, more than any other, can be tightened almost indefinitely to accommodate wood movement. The drawings on the next two pages detail the joints I've described here.

Wedged through mortise and tenon

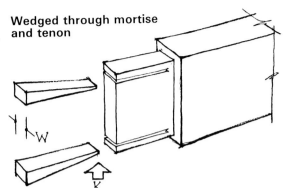

The wedged tenon may be through or blind, using one wedge at each end of the tenon, as shown here, or two wedges at each end. Wedges are frequently made out of a different wood than used for the main work. If the color contrast is dramatic, accuracy at each stage is all the more necessary, for when the work is cleaned and polished, a narrow band of wedge at one end and a wide band at the other transmits an extraordinarily loud message. The thickness of the saw kerf is of little importance—use a tenon saw or a dovetail saw. Saw down the tenon $\frac{1}{16}$ in. to $\frac{3}{16}$ in. from the edge—the exact distance varies according to the pliability of the wood. Stop the cut $\frac{1}{8}$ in. to $\frac{1}{4}$ in. from the shoulder, to give the wood a better chance to bend.

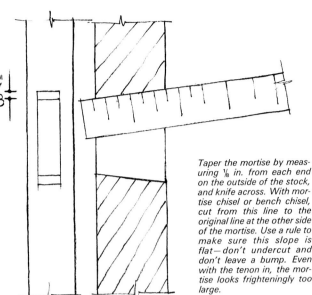

Taper the mortise by measuring $\frac{1}{8}$ in. from each end on the outside of the stock, and knife across. With mortise chisel or bench chisel, cut from this line to the original line at the other side of the mortise. Use a rule to make sure this slope is flat—don't undercut and don't leave a bump. Even with the tenon in, the mortise looks frighteningly too large.

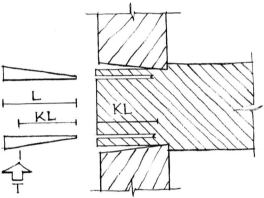

KL is the length of the saw kerf.

K is wedge thickness at tip, which is thickness of the saw kerf.

L is the length of saw kerf plus $\frac{3}{8}$ in.

W is width, same as width of mortise.

T is the thickness at length KL, which is the tip thickness K plus the extra opening at the outside of the mortise, usually $\frac{1}{8}$ in.

To make wedges, prepare stock with grain lengthwise, as thick as the mortise width. Square one end and square round at length L. Also square round the wood at length of kerf KL. Measure kerf thickness K from the edge at the top, square across. On line KL measure in from the edge T. Connect points and saw down the outside of this line. To make the second wedge, measure K along line L from the edge of the first kerf, and with try square mark to the end of the wood. Saw the outside of this line and proceed as before.

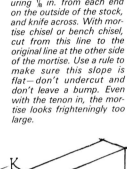

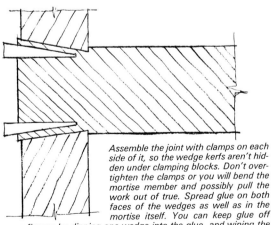

Assemble the joint with clamps on each side of it, so the wedge kerfs aren't hidden under clamping blocks. Don't overtighten the clamps or you will bend the mortise member and possibly pull the work out of true. Spread glue on both faces of the wedges as well as in the mortise itself. You can keep glue off your fingers by dipping one wedge into the glue, and wiping the excess onto the other. Drive the wedges with an iron hammer, not a mallet—you'll hear a distinct change in the sound of the blows as the wedge fills its space and becomes solid. Hit them alternately so they enter together. Once the wedges are snug, if time is pressing, you can remove the clamps and continue working. But if you are using a C-clamp across the joint to get a good cheek-to-tenon interface, best leave it alone until the glue has cured.

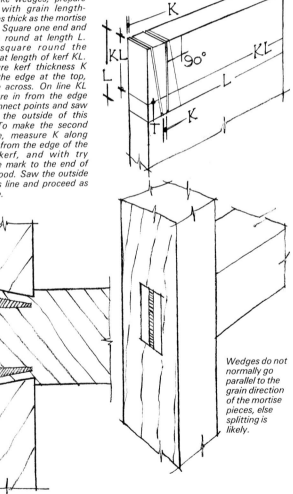

Wedges do not normally go parallel to the grain direction of the mortise pieces, else splitting is likely.

The blind version of this joint—foxtail wedges—requires strict attention to accurate measurement. Make the wedge without any excess wood on the end, and when cutting the mortise leave a good pad of wood on the blind side—at least $\frac{3}{8}$ in. Clamp with a support block on the blind side, since the wedges press hard on the bottom of the mortise and are liable to damage the wood.

Draw-pegging

Choose a dense and somewhat pliable material for draw pegs and consider the diameter. Large pegs in buildings—treenails—are usually oak, hickory or ash. These ring-porous woods are less suitable for small work, where maple, birch or a denser exotic wood are preferred. Position is also a matter for judgment: If too close to the shoulder line, the mortise cheeks might split. If too close to the end of the tenon, the wood may fail in shear along the grain, on lines SL. If distance BC is less than 1½ in., use only one peg.

In old work the pegs were often square, hexagonal or octagonal in section, and split out rather than sawn. Their corners bite into the drilled hole, for increased holding power. Their heads are often left proud, as a design detail, chamfered or whittled.

Make the mortise-and-tenon joint in the usual way. Whether one or two pegs are used depends on the size and shape of the members. Fit the joint, remove the tenon and drill through both cheeks of the mortise at Y. Replace the tenon, clamp the joint, and mark the center of the hole on the face of the tenon. This can be done by putting the drill back in the hole and giving it the lightest turn. Remove the tenon again, and drill the same sized hole about $\frac{1}{32}$ in. nearer the shoulder line from the marked center, at X. Thus the two holes are slightly off center. When the pegs are driven home the joint will be drawn tight. The procedure may be done entirely by measurement.

The tenons in this type of cross-joint may be short—place the pegs with care.

Tusk tenons

The wedged tusk tenon, left and right, is made in the same way as a through tenon. It can run vertically or horizontally—there is no formula. These drawings are only explanatory, and by no means exhaustive. Keep these points in mind:
1) Make the projecting tenon considerably longer than the thickness of the mortise piece, to guard against splitting along the grain when the wedge is driven home.
2) Make the wedges long enough for a good bearing surface on each side of the mortise. A short wedge will crush the fibers of both pieces.
3) Leave a shoulder all around the tenon piece, also for a good bearing surface.
4) Angle the wedge on one side, but not too much or it will loosen itself. A gradient of one in six is generally satisfactory.
5) Make the angle on the outside of the wedge mortise match the angle of the wedge itself.
6) About ⅛ in. of wedge hole extends inside the mortise piece—else the joint cannot tighten.

One variation is to use two similar wedges (folding wedges). The hole for the wedges is square at both ends, but it still must extend into the mortise piece to allow the joint to tighten. Drive both wedges home at once and persuade someone to hold protective blocks on their ends to avoid the damage of a glancing blow. To reduce this risk, angle the ends of the wedges and use a hammer, not a mallet.
To make the projecting tenon shorter and reduce the risk of shear, a slot may be drawn down its center and a key or spline of wood glued in, cross-grain. This effectively makes the tenon into a form of plywood. If the wedges become loose as a result of seasonal humidity changes, it is a simple task to tighten them.

The Haunched Mortise and Tenon
How to strengthen the corner joint

by Ian J. Kirby

The most basic of woodworking problems is joining two pieces of wood together at right angles to form a corner. The most common joint for doing this is the mortise and tenon. We are usually in one of two situations: first, where two pieces of similar thickness are being joined, as in the corner of a door frame (fig. 1); and second, where a third piece of wood of dissimilar thickness is involved, as in a typical table or chair joint between apron and leg (fig. 2).

When designing these joints, note that the top edge of the mortised member—the vertical piece in the illustrations—will be in line with the top edge of the tenoned rail. If you want the final appearance to be as shown, then the joint must be stopped somewhere below the top surface. The usual solution is to add a haunch, which may be either square or sloping. Both variations strengthen the joint and increase the gluing area, and the basis for choice is visual. If you want a clean, uninterrupted line at the top shoulder of the joint, you would use the sloping haunch. If you don't mind the interrupted line or if the joint will be concealed, the square haunch is a little easier to make and a little stronger.

The illustrations show the form of the joints, and the elevations suggest suitable proportions. I must emphasize here that it is the responsibility of the designer to detail all the joint dimensions to achieve the visual effect he wants as well as the mechanical strength the structure requires.

The main reason for the haunch is strength. Resistance to twist is especially improved. If you leave the haunch off altogether (fig. 3), the result is that about a third of the width of the rail is free-floating, with no mechanical bond and no glue bond where it needs it the most. If you go to the other extreme and make a bridle joint, you sacrifice mechanical resistance to downward loading. Although glue is strong in shear

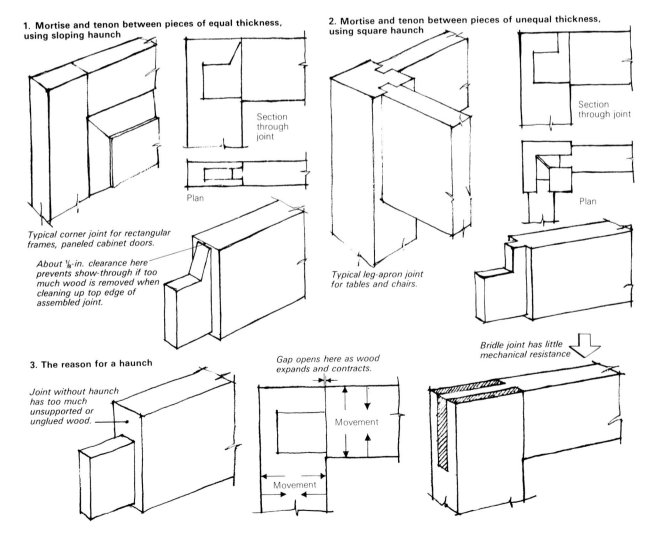

1. Mortise and tenon between pieces of equal thickness, using sloping haunch

Section through joint

Plan

Typical corner joint for rectangular frames, paneled cabinet doors.

About ⅛-in. clearance here prevents show-through if too much wood is removed when cleaning up top edge of assembled joint.

2. Mortise and tenon between pieces of unequal thickness, using square haunch

Section through joint

Plan

Typical leg-apron joint for tables and chairs.

Bridle joint has little mechanical resistance

3. The reason for a haunch

Joint without haunch has too much unsupported or unglued wood.

Gap opens here as wood expands and contracts.

Movement

Movement

Illustrations: Ian J. Kirby

and the joint has a great deal of gluing area, the two parts meet at right angles. This puts considerable strain on the glue line when the wood shrinks and expands, and the condition is aggravated by the large amount of exposed end grain. In addition to the forces the object will encounter in use, you must consider seasonal shrinkage and expansion. These can exert far greater forces than ordinary usage will, although they do not occur or show themselves for some time. The most common fault in this regard is a gap appearing at the top edge shoulder line.

There is a peculiarity of manufacture common to both the sloping and square versions of the haunched joint. When cutting the wood to length, an extra ¾ in. should be left on the end of each piece where a mortise will be made. The wood should be knifed quite deeply all around at the true length where the top edge of the tenoned piece will be aligned. The extra ¾ in. of wood is left there until the joint is made and the glue is cured, whereupon it is sawn off. The extra wood, called a horn (fig. 4a), has the effect of making the mortise be more in the center of the piece of wood than right at the end. While the joint is being made, it helps prevent the wood tissue from splitting beyond the joint. While the joint is being glued and clamped together, provided it fits well, the piston-and-cylinder action can very easily crack the wood in the area of the haunch. It is also possible, especially on a blind joint, for a piece of short grain to pop right out along-side the haunch. The horn prevents these unhappy events. Since glue seals the piston-and-cylinder, the viscosity of the glue is also a factor. The thicker it is, the more slowly you should clamp the joint together—the excess glue has to go somewhere, and it needs a little time to flow out of the joint. Imagine the dilemma in the days of hot animal glue: As the glue cools, it becomes more viscous and starts to set, so clamping slowly doesn't work. Part of the solution then was the horn, and it still is.

The square-haunched mortise and tenon is marked out as shown (figs. 4b, 4c). A longer haunch would seem to give more gluing area, but it might also allow the mortise cheeks to curl away from the tenon. A haunch shorter than square gives too little gluing area. The tenon is sawn in the usual way, being sure to leave the shoulder lines for last.

The mortise is chopped as usual, but only in the full-depth part of the joint. Remove the waste to accommodate the

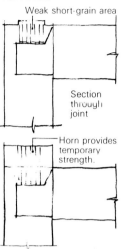

Short grain on mortise member is quite weak until joint is glued—top of joint may be pushed out if no horn is left.

Weak short-grain area

Section through joint

Horn provides temporary strength.

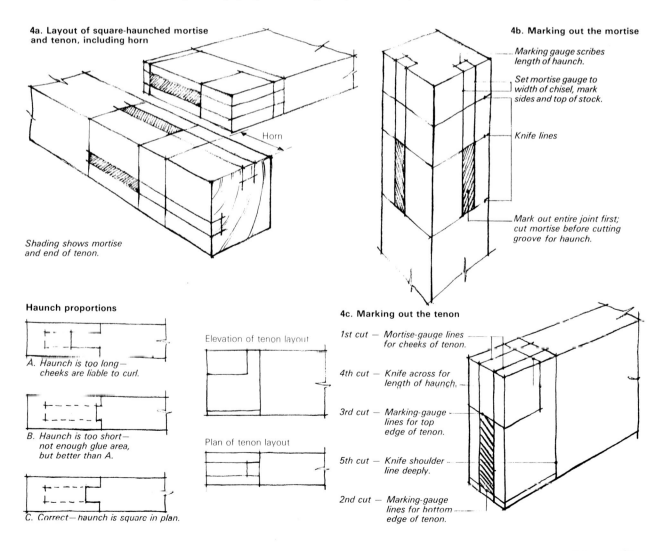

4a. Layout of square-haunched mortise and tenon, including horn

Horn

Shading shows mortise and end of tenon.

Haunch proportions

A. Haunch is too long— cheeks are liable to curl.

B. Haunch is too short— not enough glue area, but better than A.

C. Correct—haunch is square in plan.

Elevation of tenon layout

Plan of tenon layout

4b. Marking out the mortise

Marking gauge scribes length of haunch.

Set mortise gauge to width of chisel, mark sides and top of stock.

Knife lines

Mark out entire joint first; cut mortise before cutting groove for haunch.

4c. Marking out the tenon

1st cut — Mortise-gauge lines for cheeks of tenon.

4th cut — Knife across for length of haunch.

3rd cut — Marking-gauge lines for top edge of tenon.

5th cut — Knife shoulder line deeply.

2nd cut — Marking-gauge lines for bottom edge of tenon.

haunch by placing the workpiece in the vise at a slight angle and cut down the insides of the haunch lines with the tenon saw (fig. 5). Only a few inches of the saw can be used, but the method is fast and efficient. Sight across the back of the saw to make sure it is parallel to the top edge of the wood, and be sure that sharpening has not moved the saw's teeth out of parallel with its back. Once the sides of the haunch groove are sawn, remove the waste with the mortise chisel. Place it about halfway down the haunch groove on the end grain and give it a smart tap. Watch out for grain direction, in case it is running down into the joint and liable to result in more waste removed at the joint end than at the horn end. With care, it is not difficult to keep the haunch groove clean and parallel. Two points: The groove has to be cut in the horn, which seems a waste of time, but that's how it is. Second, don't try to chop the haunch groove the same way as the mortise. It is difficult this way to keep the bottom clean and parallel.

The sloping haunch joint is marked out in the same way as the square haunch (fig. 6). At its root, the haunch is usually as long as the tenon is wide—if the joint were sectioned here, it would look in plan the same as the square haunch. Saw the tenon cheeks and ends in the usual way, then saw the slope before sawing the shoulders. A common error is to saw the main tenon and shoulders before the sloping haunch, thereby removing the layout lines.

The full-depth portion of the mortise is chopped as usual. Then place the wood in the vise and cut the slope with the mortise chisel. There is no measurement one can make to get the slope right the first time. The normal method is to cut the slope short, check the depth at its root and remove the necessary amount. It is very easy to draw the joint showing the slope of the haunch coming right to the top of the shoulder line. But it should stop short, leaving about ⅛ in. of vertical shoulder before the slope begins. This allows for over-enthusiasm when planing the top edge clean. □

Louvered Doors
Router jig cuts slots

by William F. Reynolds

Woodworkers at the U. S. Capitol are often called on to match the decor of an earlier age. When the Architect of the Capitol decides that a room must be renovated or redecorated, woodworkers like Ned Spangler, a cabinetmaker for the U. S. Congress, must rise to the occasion and improvise techniques to carry out the project. An example of the challenge, and the solution, is the production of louvered doors. Before air conditioning, Washington offices were extremely humid, and louvers allowed air circulation into cabinets, which kept the contents from mildewing and prevented doors from sticking. Although louvers can be difficult and time-consuming to make, they are elegant and dignified.

Spangler has a shortcut for making the louvered doors he is so often required to build. As Spangler says, "The job has to be done right, and right the first time." His router jig turns out slotted stiles quickly and precisely. It could be adapted to make a series of mortises, such as for crib slats.

The jig consists of two long hardwood posts mounted on plywood, a square piece of Masonite screwed to the baseplate of the router, and two router carriages that slide along the posts and guide the router as it cuts the equally spaced, mirror-image slots on opposite stiles. The jig shown here is designed for a ¼-in. router bit, and for ⅞-in. thick stiles for a

5. Saw down the walls of the haunch...

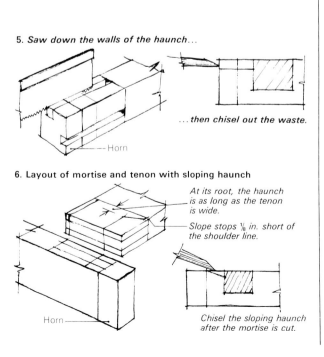

...then chisel out the waste.

Horn

6. Layout of mortise and tenon with sloping haunch

At its root, the haunch is as long as the tenon is wide.

Slope stops ⅛ in. short of the shoulder line.

Horn

Chisel the sloping haunch after the mortise is cut.

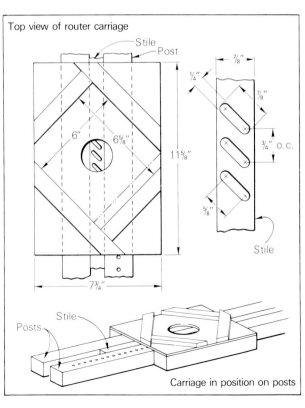

Top view of router carriage

Carriage in position on posts

Test slots are checked for accuracy of angle, fit.

Spangler routs a slot; screwdriver indexes carriage.

The completed door—elegant and dignified.

30-in. door, with 1-in. wide slats angled 45° and spaced ¾ in. on center along the stiles.

The posts, cut from birch or similar hardwood, should be long enough to hold the stiles firmly, with perhaps an extra foot on each end. For this jig, the posts are 5 ft. long and 1¾ in. square. They are mounted parallel to each other and ⅞ in. apart, on a plywood base, so that a stile fits snugly between them.

To mount the router, cut a 6-in. square from ¼-in. Masonite. In it drill three holes to match those for screws in the router base, and one of ¾-in. dia. in the center, for router bit clearance. Then screw the square to the base of the router.

Next, drill holes in the posts for stops for the router carriages. These equally spaced holes determine the spacing of the slots, and therefore of the louvers. Drill 1-in. deep holes, ¼ in. in diameter, down the top of either post, starting and ending 8 in. from each end. Space the holes for minimum clearance between louvers, in this case ¾ in. on center.

The two router carriages slide along the posts. At each stop, they allow the router to travel the exact length of the 45° slot it must make for the louver. To make these carriages, cut two pieces /¾ in. by 11⅝ in. from ¼-in. plywood and lay out a 2-in. diameter hole at the center of each. These holes will help align the louver slots and allow clearance for the router bit. Cut four pieces, 2 in. by 1¾ in. by 11⅝ in. Nail two to the underside of each plywood plate, to form the carriage sides and keep it centered as it moves along the posts.

Four pieces, each 1 in. wide by ½ in. thick, position the router base plate atop the carriage and allow it to travel only far enough to cut a slot. Starting from the center of the uncut hole, measure to each side and lay out a rectangle 6⅝ in. by

6 in. at 45° to the sides of the plywood. The router base will travel within this rectangle. For the second carriage, the long sides of the rectangle should slope 45° to the other side; thus the two will produce mirror-image slots. Make sure that this is so before nailing the pieces in place to frame the rectangles. Then cut the 2-in. center holes. With this setup, the router travels ⅝ in. to make a slot ¼ in. by ⅞ in. long. If you use a bit larger than ¼ in., adjust the router travel distance to the length of the slot minus the diameter of the router bit.

To cut the first slot, set one stile between the posts and place the carriage on top. Mark the stile where you want the cut to end, perhaps an inch from the end of the stile, with the router travel from lower right to upper left. Stand a sample slat, cut short, on the stile inside the hole in the jig and align the sample so it marks the area for the first slot. Put a ¼ in. screwdriver in the post hole nearest the point where it will hold the jig in place while the slot is being cut. It may be necessary to move the stile slightly, leaving the screwdriver in the selected hole, to achieve perfect alignment.

Set the router bit to make a test cut, on this or on a test stile. Cut to the required depth, usually ⅜ in. to ½ in. Successive cuts are located by advancing the screwdriver to the next post hole and moving the carriage along.

To cut the slots in the opposite stile, use the second carriage. Always make a trial cut to check for alignment. If the first slot matches the one on the mating stile, then the others will too. If you have measured exactly, the job should go quickly and the louvers should fit the first time. ☐

Bill Reynolds, a dedicated amateur woodworker, is a freelance journalist based in Washington, DC.

Mortise & Tenon by Machine
With help from jigs and fences

by Ian J. Kirby

Woodworkers have devised endless methods for cutting mortise and tenon joints, relying upon hand tools, machine tools and various combinations of the two. Deciding which method to use depends primarily on the tools one has at one's disposal.

The articles beginning on pages 6, 11 and 27 focused on hand-tool methods, which have several virtues. The tools are not special. There is a logic to the process. It is reasonably quick. Once one has mastered the skill, one can achieve the desired result exactly. Having designed a joint, the workman need never compromise in its manufacture. However, the result is always at risk and one must concentrate to avoid spoiling it. It can become exceedingly tedious if one has a lot of joints to do.

Special-purpose machines designed for mortising are one alternative. I'll discuss some of them later; they are generally fast and accurate, but expensive and beyond the needs of most shops. The middle ground is to use a machine not specifically designed to cut a given joint, such as the table saw, radial saw, drill press or router. These machines, with the assistance of suitable jigs, can remove the bulk of the waste accurately and efficiently. Some hand-finishing can then produce the desired result. The notion that there is only one way to achieve a result is simply wrong, for every workman develops his own techniques, and this article thus cannot be exhaustive. But regardless of methods, every workman should arrive at the same result in the end. The available tools do not determine the size or proportions of the joint, nor excuse inaccuracy in its manufacture.

The mortise — To deal with the mortise first, and ignoring such details as sloping haunches, twin joints and dimensioning, the main consideration is that the two inside faces be parallel to each other and to the face side or edge of the stock.

A frequent question is, "Can I drill out most of the waste and then pare down the cheeks with a wide chisel?" The answer is "yes" to the drilling, and "no—or only with great difficulty" to the chiseling. To keep the mortise square and parallel when using a wide chisel really requires a jig. Sighting the chisel while paring across the grain is too hit-or-miss. And a jig would probably be too complex because of the nature of the operation. An acceptable result can be achieved, however, by drilling a row of overlapping but undersized holes to remove the bulk of the waste, and squaring up to the line with a mortise chisel. The joint still has to be marked out with the mortise gauge, to assist at the chiseling stage.

In any machine operation, one must think of the cutting tool itself in close association with the fences and guides related to it. Usually there is no marking out for a joint made entirely by machine. Layout marks tell where to cut a joint—with a jigged system one wants to be forced to cut where the jig directs. When using a drill press (not a portable drill) to remove the waste, the machine's depth stop establishes the mortise depth. A fence fixed to the table so the face side or

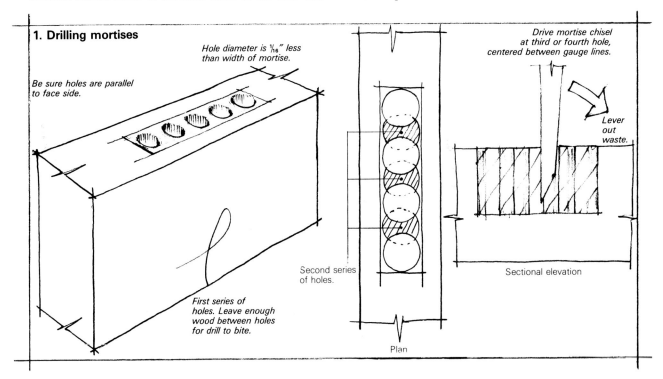

1. Drilling mortises

Hole diameter is ¹⁄₁₆″ less than width of mortise.

Be sure holes are parallel to face side.

First series of holes. Leave enough wood between holes for drill to bite.

Second series of holes.

Plan

Drive mortise chisel at third or fourth hole, centered between gauge lines.

Lever out waste.

Sectional elevation

A

B

C

D

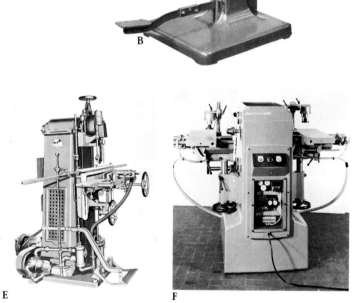

E   F

A. *Hollow-chisel mortising attachment for drill press (Rockwell model 15-840). Y-shaped yoke at top attaches hollow chisel to quill; lower yoke is part of fence arrangement clamped to drill table.*

B. *Hollow-chisel mortiser, Oliver No. 194. Table includes hand wheel for clamping work against fence, with hold-down mounted on vertical column. Wheels below table control sideways travel, tilt and height. Foot pedal moves chisel and motor assembly into the work.*

C. *Inca horizontal-boring and mortising option on 10-in. table saw uses three-jaw chuck attached to saw arbor. It includes work clamps and adjustable stops. Hand wheel at bottom raises and lowers table, and levers control infeed and crossfeed.*

D. *Griggio slot mortiser, from Italy, takes end-mill cutter in stationary horizontal chuck. Hand wheel raises table, clamp holds work, levers move table and work in and out, back and forth. Sold by H. Weigand Corp., Claremont, N.H., and Carpenters Machinery Co., Philadelphia.*

E. *Chain-saw mortiser has hydraulic clamp and feed—operator loads the stock, taps the foot pedal, and unloads it. Photo: Northfield Foundry & Machine Co., Northfield, Minn.*

F. *Bacci oscillating chisel mortiser, also from Italy, has double-ended cutter shaft and two tables for production work. Cutter rotates at 8,750 RPM and also swings back and forth 200 times per minute. Pneumatic tables move synchronously in all three planes. Thus the size and shape of the mortise is virtually unlimited. Photo: Richard T. Byrnes Co., West Chester, Pa.*

edge of the wood can be placed against it establishes the distance to the center of the drilled holes. Two end stops determine the left-to-right travel of the workpiece, and if the wood is squarely placed within these fences, the correct side uppermost and the right way around, then the series of holes can be drilled only within the defined parameters

The diameter of the drill, however, should be at least ¹⁄₁₆ in. less than the width of the mortise chisel that is to be used to clean out the remainder of the waste. The pattern of holes depends in part on the type of drill bit. Best is a bit with two scribing lips, like a Russell-Jennings. A Forstner bit also gives good results. An engineer's bit for drilling metal is not so effective; a spade bit gives variable results depending on the type of wood and on the feed and speed.

Drill the end holes first, then drill along leaving up to ¼ in. between holes (figure 1). Drill out the remainder by positioning the spur on the webs of wood left between the holes. The edges of a drill may overhang, as long as the center is cutting into solid material. The drill drifts when the center is not cutting firmly into wood. Drive the mortise chisel straight

down into one of the middle holes, about half-way to full depth, and carefully lever out the small amount of tissue remaining on the walls of the mortise with the chisel's bevel downward. Don't try to go to the bottom in one cut—you'll quickly get the feel and realize that the operation can be fast and simple. Finish the ends by knifing the line and driving the chisel straight down, just as when doing it all by hand.

Mortising machines — The hollow-chisel mortiser is free-standing, with built-in table, fences, clamps and stops. In small shops it is usually an attachment for the drill press—a square chisel with a hole in the center through which an auger-type drill fits (figure 2, p. 86). A yoke fastens the chisel to the drill-press quill, so the chisel and drill will move together into the work, but only the bit rotates. As the bit removes most of the waste, the chisel, sharpened on the inside to form four cutting edges, follows to shear out the remaining wood and force it into the auger. The chisel shaft has at least one window through which chips can escape.

The quality of a hollow-chisel attachment is closely related to its price—a good set for ¼-in., ⅜-in. and ½-in. mortises will cost about $200. Before ordering any attachment, be sure it is compatible with your drill press. You don't have to worry about whether the drill press can stand the work load—it will.

When mounting the tool, make sure the plane of the table is at right angles to the bit. Also make sure that the square chisel has its inside face parallel to the fence. Adjust the bit so it does not touch the sharpened end of the chisel, otherwise both will overheat. Aim for a gap of 1/32 in., enough to loosely fit a business card. It's usually possible to jig the hollow-chisel mortiser so that you don't need any marking out on the wood (figure 2). For short runs, it's probably easier to square pencil lines across the wood and omit the end stops.

It is normal to cut the end holes first, then to cut intermediate holes with wood left between them, and finally to clean out with another pass along the work. This is because the hollow chisel tends to drift if it is not cutting on all four edges, or on two opposite edges. Each worker will find a pattern that suits him. The square chisel is reluctant to withdraw from some woods. The remedy is to polish the outside of the chisel to reduce friction, and to reach full depth by ¾-in. bites. Withdraw the bit, clear the waste, and take a second ¾ in. in the same place. This characteristic of the machine makes it imperative to hold the wood firmly down on the bed, by clamping one or more blocks onto the back fence. The fence shown in figure 2 is simple and sturdy, and worth making well since good jigs are a tooling investment.

A small but important point is to keep the whole of the fence rig clear of chips, so they don't get between the workpiece and fence. In industry, a squirt of compressed air does it. Next best is to keep a brush at the machine and sweep off the bed and jig after each cutting. Many people minimize the

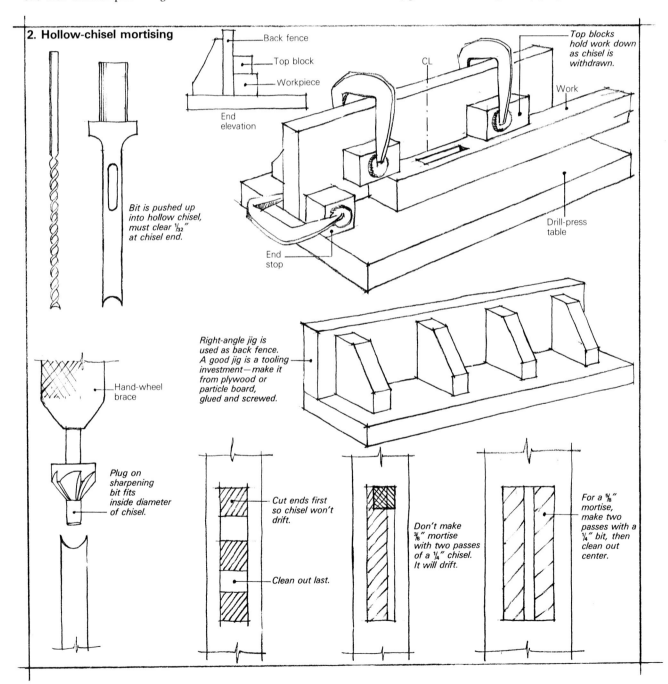

2. Hollow-chisel mortising

Back fence

Top block

Workpiece

End elevation

Bit is pushed up into hollow chisel, must clear 1/32" at chisel end.

Hand-wheel brace

Plug on sharpening bit fits inside diameter of chisel.

Top blocks hold work down as chisel is withdrawn.

CL

Work

End stop

Drill-press table

Right-angle jig is used as back fence. A good jig is a tooling investment—make it from plywood or particle board, glued and screwed.

Cut ends first so chisel won't drift.

Clean out last.

Don't make ⅜" mortise with two passes of a ¼" chisel. It will drift.

For a ⅝" mortise, make two passes with a ¼" bit, then clean out center.

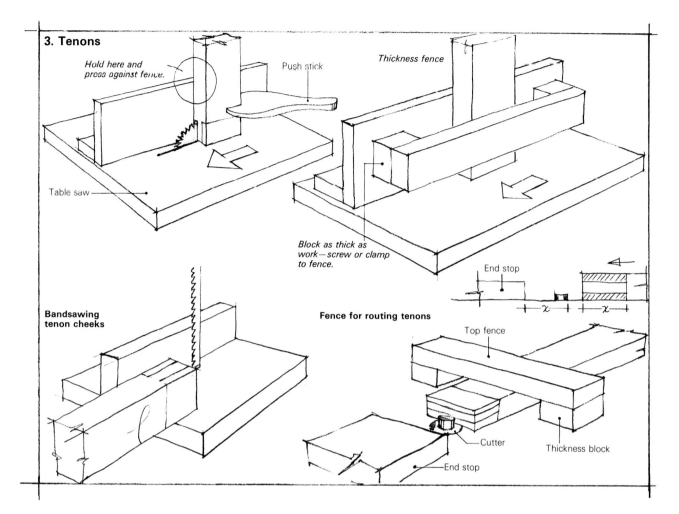

3. Tenons

Hold here and press against fence.

Push stick

Table saw

Thickness fence

Block as thick as work—screw or clamp to fence.

End stop

Bandsawing tenon cheeks

Fence for routing tenons

Top fence

Cutter

Thickness block

End stop

problem by cutting grooves and reliefs along the inside corners of jigs and fences, but the brush is still necessary.

It is not good practice to make a ⅜-in. wide mortise with two passes of a ¼-in. bit. The bit will be cutting on only three sides during the second pass, and it will probably drift. The ¼-in. bit will make a ⅝-in. mortise, via two passes on each side and a third down the center.

A hollow chisel is sharpened with a bit that looks like a rose countersink with a cylindrical plug on the end. The outside diameter of the plug is a hair smaller than the inside diameter of the chisel. Set the chisel upright in a vise, load the bit into a wheel brace, and place the plug into the chisel's bore. The reamer flutes are very effective and only a few turns of the drill with light pressure will remove enough metal. Don't use an electric drill for sharpening—it goes too fast and you can't feel the action. Sharpen the drill bit in the usual way, from the inside of the angle so its diameter doesn't change.

Among the more specific machines for mortising is the horizontal slot mortiser or long-hole borer. Like the router, it leaves a round end. Fundamentally, it consists of an end-mill style cutter, with a sharpened end and sharp flutes, revolving horizontally over a traveling bed much like the cross-slide of a metal lathe. The bed moves the work into the cutter to full depth, then traverses to make a mortise. The same machine can also make tenons, forming one cheek and shoulder with each horizontal traverse. Some versions of this machine hold the work stationary and move the cutter into it.

A chain mortiser is akin to a chain saw, with its bar held

vertically and set into a slide device. The system is not used much in the furniture industry, being better for long and deep mortises in large-sectioned material such as fence posts.

Probably the most sophisticated mortising machines use a swinging and orbiting cutter, driven by a cam system and a little like a sewing machine writ large. These machines can cut an absolutely accurate mortise through any kind of wood, even plywood, without regard for knots or end grain.

Tenons — Generally, making tenons with a nonspecific machine is not as difficult as mortising—although there are probably as many variations on the theme. The most common tool for cutting the cheeks is the table saw, with the work held vertically by a fence and passed over the blade. A carbide-tipped blade gives best results. If the tenon is centered, both cheeks can be cut at the one setting. You can build or buy a suitable fence that rides in the crosscutting slots. A stationary fence must have enough overhang before and after the sawblade to support the full width of any piece being tenoned, and it is probably best to make it the full width of the table. Using the simple fence shown (figure 3), the left hand (assuming you are right-handed) holds the top of the workpiece against the fence. At the same time, with the aid of a push-stick, the right hand traps the work and pushes it forward in the direction of the arrow. If you do not feel completely safe using this method, then arrange a thickness fence as well.

The simplest way to cut the shoulder, of course, is to saw it

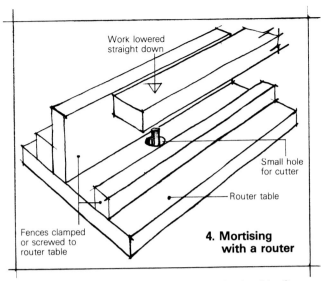

4. Mortising with a router

Labels in figure:
- Work lowered straight down
- Small hole for cutter
- Router table
- Fences clamped or screwed to router table

with a transverse fence. Whether you cut the shoulder first or the cheek first is a matter of personal style and there are arguments both ways. I prefer to cut the cheek first in order to have the largest bearing surface while the work stands on end. Also, if the shoulder has already been cut, the small block of scrap often wants to get back on the saw teeth and then fly around the room in an unsettling way. If one has a dimension saw—that is, one with a traveling table—the same fence system can be mounted right on it. The workpiece can then be clamped onto the vertical fence, and both hands can be employed pushing the bed and workpiece past the blade.

The slowest method is to set the saw for the shoulder cut and move the work along one kerf thickness at a time. It may be the handiest method with a radial arm saw, however, and the operation speeds up considerably when one substitutes a dado head. Most radial arm saws can be rotated through 90° and locked parallel to the table surface, whereupon the wood can be laid flat on the table and the saw pulled through the cheek. The work requires a platform and fence, plus room to clamp each piece in place, as the amount of outward thrust can be considerable. The method is efficient for quantities of identical parts.

When used to cut tenons, the band saw could be considered an automated backsaw, since it is easy to use freehand. The wood must be truly square and the blade running perpendicular to the table, otherwise the tenon will come out at some odd angle. This is workmanship of risk, and the decisions about where to cut and where to stop are no different than when working by hand. It is feasible to use a fence with the band saw, although the blade has to be sharp and tensioned just right, else it will wander. Some band saws just do not seem to have the capacity to saw a straight line when using a fence, no matter what one does to try. Feed speed and the hardness of the wood are contributing factors (see p. 96), and generally a slow feed gives the best results.

Because the waste being removed is shallow, a router can mill a good tenon. In most cases it is best to cut the shoulder lines with the radial arm or table saw first. Then lay the work flat on the router table and pass it over a straight cutter or any end-mill style cutter. Fences are as necessary here as with any other method, although the cut is easy and it is tempting to wing it. Please don't take the chance.

There are also a wide variety of industrial tenoning machines, many of them using shaper-style cutterheads mounted in pairs to mill both faces of the tenon at once. Such machines are very efficient and suitable only for high-volume work.

The router — The electric router, combined with a careful system of jigs and fences, is a useful mortiser. When making a wide mortise—anything over ½ in.—it's better to make two slots ¼ in. wide at each side, and then to remove the waste from the middle. When making a deep mortise, go to the depth in two or more bites.

Don't try to drop the router into the work. Use a table with the router hung underneath, so the bit will project upward through its surface. The minimum number of fences is two, one along each side of the work. End stops are always a help, but their use is often limited by the size of the router table. The common method is to rest one end of the work on the table and lower the other end onto the cutter. This is not the best way, since the work comes down in an arc. Instead, hold the work parallel to the table against the fence (figure 4), and lower it straight down with both hands, keeping the hands well away from the cutter. The dimensions of the work will dictate the dimensions of the fences, but the aim is to arrange the system so you can keep a tight hold on the workpiece. We usually make fences long-grained in the direction of travel, and if the workpiece projects well above the fences, control is not hard to achieve. But if the workpiece is small in cross section, make the fences by simply clamping wide boards, cut off square, flat on the table.

Most problems in routing mortises arise because of the small size or make-do nature of the router table. It's worth investing in a piece of coreboard or good-quality plywood and making a proper large table, once and for all. Rout a recess underneath so you don't lose the table thickness from the depth of cut, and keep the cutter hole small so the table will support the wood right up to the cut. The larger the hole, the harder it is to measure and to visually assess cutter height, and the easier it is for fingers to get into the hole.

It also pays either to devise adjustable fences for the table, or else to screw the fences down. It may seem odd to go to some trouble to get a clean, often expensive working surface only to mar it with screw holes, but the life of the surface will be longer than you imagine. If you keep the fences with screw holes already drilled in them, the system is easy to use and quickly set up. It's usually safer and more accurate than clamping down whatever comes out of the short-ends box.

A routed mortise has round ends. One can shape the tenon to match, or one can finish the mortise square with the conventional chisel. There are points on both sides. When the mortise goes through, round ends make a most acceptable design detail. It takes some skill to round the tenons, but it's worth the effort. An alternative is to leave the mortise round and the tenon square, and to force the two together. This procedure is not unusual in production. The width of the tenon is made so that its corners will bite into the semicircular ends of the mortise. The crushed corners create a tight friction fit, enabling the assembly to be taken from the clamps after a very short pressure time. □

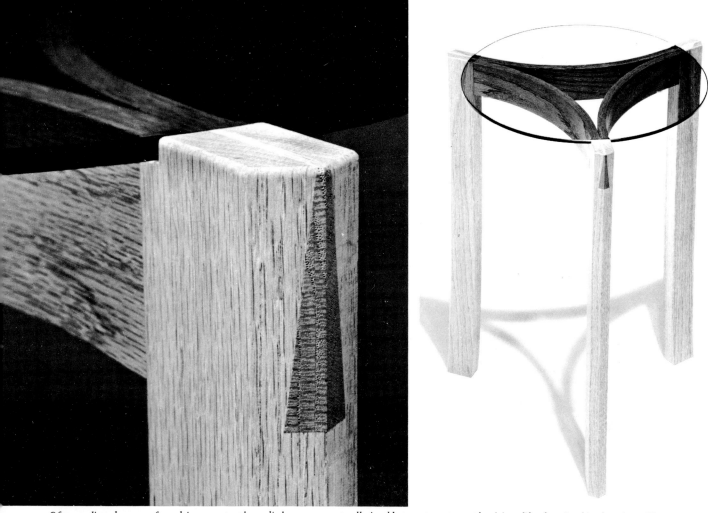

Often maligned as unsafe and inaccurate, the radial-arm saw actually is able to cut neat, complex joints like those in this three-leg table. Erpelding cut the dovetail mortises in the legs as shown on page 42. The tenons on the triangular stretcher assembly, composed of bent laminations, were scribed from the mortises, then cut with a back saw—an efficient interplay of hand and power tools. Photos: Steve Young.

Slip Joints on the Radial-Arm Saw
Getting accurate results from a versatile machine

by Curtis Erpelding

Most discussions of power-tool joinery focus on the table saw, the router and the bandsaw, and neglect the radial-arm saw. Many insist that the radial-arm saw is inherently dangerous, and impossible to adjust accurately. I find neither argument convincing. Caution and common sense go a long way in reducing potential hazard, and the saw's ability to do precise work is almost completely in the control of the operator. My saw is a mid-60s vintage 10-in. Sears Craftsman. The guides for the rollers are worn and pitted, the indexing pins less than tight, and the arbor has a slight amount of run-out. Nevertheless, I've been using this saw for five years to cut close-fitting joints, and would hesitate to replace it with a new industrial model. As a good marksman knows how to deal with the idiosyncracies of his rifle to group his shots, so the canny woodworker knows how to compensate for slop and play in his radial-arm saw.

In making slip joints, the radial-arm saw offers several advantages over the table saw. First, the workpiece remains fixed during the cut, and because there is less slop in an old radial-arm saw than in most new tenoning jigs (homemade or otherwise), the cuts are more precise and easier to control. Second, the column-raising mechanism on the radial-arm saw lets you make very fine depth-of-cut adjustments for paring a joint to final fit. You'd have a hard time adjusting the fence on a table saw in such small increments to make the same cuts. Finally, the radial-arm saw imposes no real limits on the length of the pieces being joined, as long as the ends are supported. With the table saw, cutting joints on members more than 4 ft. long involves a precarious balancing act, which will adversely affect the accuracy of the joint, particularly if the free end runs afoul of a ceiling joist or light fixture in a shop with a low ceiling.

Cutting a slip-joint mortise and tenon on the radial saw requires, as in hand work, careful marking out. But you need to mark out only one joint, which becomes the reference from which the saw is set to cut the others. After the members are

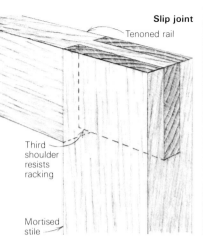

Slip joint

Tenoned rail

Third shoulder resists racking

Mortised stile

A

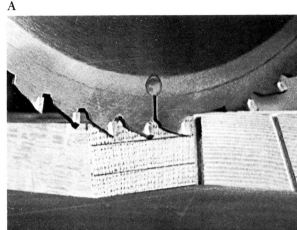

B

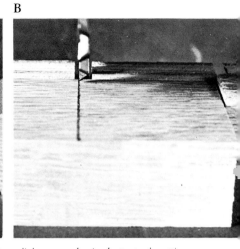

Cutting slip-joint tenons on the radial-arm saw begins by properly setting the blade to cut the shoulder line. Set the depth of cut by aligning the blade with the marking-gauge line that delineates the tenon thickness on the end grain (A). Then align a tooth that has outside set with the shoulder line (B). Sawing the cheeks requires rotating the blade to its horizontal position and positioning the stock using a 7-in. high substitute fence, a spacer and an auxiliary table (C). For safety, make several passes, advancing the work into the path of the blade between passes. The waste piece is left attached and broken off by hand before the cut is completed (D).

D

E

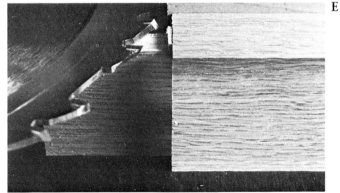

To complete the tenon, a third shoulder (E) adds, resistance to racking and hides imperfections in the mortise. Author cuts mortise for slip-joint with saw in horizontal position, using substitute fence, spacer, and auxiliary table, as in cutting tenon cheeks. A dowel in a thin piece of wood (F) stops the workpiece to control the depth of the mortise.

C

F

Photos, except where noted: John Switten

thicknessed and cut to length, they are positioned in what will be their final orientation. To avoid confusion later, mark each piece for placement and position now.

The tenons—The tenoned members (rails) should be marked first. For the shoulder lines, set the cutting gauge to the width of the mortised member and score the lines on both faces and on the inside edge as well. With a marking gauge, mark out the tenon thickness on the end grain, letting the gauge ride against one cheek and then the other. This ensures a perfectly centered joint. Where the tenon must be offset, use a mortise gauge and set the two points to the thickness of the tenon, then set the fence the appropriate distance from the face side. When the rail has been marked for the shoulders and for the tenon thickness, take a sharp pencil and make the lines more visible.

Cut the shoulders to the appropriate depth first. Set the height of the saw so that the teeth just touch the top mark on the end of the piece (photo **A**). Then line up the piece so that a tooth with outside set just touches the inside of the shoulder line (**B**). Clamp a block on the fence to hold this setting. Cut the shoulders, first one side and then, by flipping the piece over, the other.

Of all the operations, cutting the shoulders square requires the most accuracy on the part of the saw. And adjusting it to cut consistently square to the fence seems to be the major bugbear concerning the tool. If you simply can't get your machine to cut square, don't despair. Cut 1/32 in. shy of the shoulder, and then trim to the line with a rabbet plane.

But before you decide that your saw won't cut square, make all the usual initial adjustments to align it, and then practice pulling the blade into a piece of scrap that's been scored a number of times at 90° to the fence. Vary the way you pull the saw carriage over the work, observing the results of each cut. Because the blade will respond to very slight lateral pressure on the carriage handle, it could be your operating habits that keep the tool from cutting square consistently. Once you find the proper stance, the arm and the shoulder movements that make for repeated square cuts, practice until you can get it right every time.

To cut the cheeks of the tenon, turn the saw to its horizontal position. Remove the regular fence and replace it with a 7-in. high auxiliary fence on the right-hand side of the blade. This is adjusted square to the table, and so that the sawblade just touches its edge. An auxiliary table is necessary to elevate the workpiece. I use a piece from an old solid-core door; it remains reasonably stable throughout seasonal changes. Finally, clamp a block of 8/4 maple to the fence as a spacer so that its edge just touches the blade when the saw is pulled through (**C**). The spacer prevents the blade from contacting the workpiece when the saw is fully retracted.

Set the blade about 1/32 in. above the mark for the tenon. Advance the workpiece about an inch at a time, and pull the saw through. Don't try to make the whole cut in one pass or try to feed the workpiece into the saw as it cuts, or you'll overload the motor and risk a dangerous kickback. Also, don't complete the cut at this point, but leave the waste slightly attached (**D**). Remove the workpiece and break the waste off by hand. Then make the final pass, setting the shoulder line even with the edge of the maple spacer block.

Using even a 40-tooth carbide-tipped blade, cutting into tough end grain causes the blade to vibrate and leaves a cut

that's less than clean and accurate. So cut all the tenons first to this 1/32-in. margin; then trim them later with a final light cut to the gauge line, which you can make in a single pass. In operation, hold the workpiece firmly with your right hand at a safe distance from the blade. When guiding the saw carriage into the cut, keep your elbow stiff and pull by pivoting your upper body rather than by bending your arm. This helps keep the saw from self-feeding and stalling.

Though not a common practice in making slot mortises, cutting a third shoulder on the underside of the tenoned member will add strength to the joint, making it more resistant to racking. Also, the third shoulder will hide any imperfection on the inside bottom of the mortise. All that's involved is taking a slice off the bottom edge of each tenon (**E**). Set the saw to cut 1/8 in. into the tenon. Set the shoulder slightly behind the edge of the spacer block so the blade cuts just shy of the line, then chisel the remaining end grain flush with the shoulder.

The mortise—The next step is cutting the mortise. When both pieces are the same thickness, the width of the mortise can be marked out with the marking gauge at the same setting used for the tenon. The depth of the mortise is gauged from the width of the tenon and a line is scored on each edge with a knife. Continue the line across the inside and outside edges and make a slight notch with a knife on the corner edge. This notch, when lined up with the edge of the spacer block, determines the depth of cut. A long, thin piece of wood with a dowel sticking vertically out of it is clamped to the tabletop as a depth stop. With the workpiece resting against the upright dowel, there is clearance for the saw to pass (**F**). Set the stop so that the saw cuts just shy of the gauge line.

Lower the saw to cut 1/32 in. above the bottom line of the mortise layout. As in cutting the tenon, advance the workpiece only an inch or less with each pull of the saw. Make the final cut with the workpiece registered against the dowel of the stop block. Flip the piece over and repeat the process. On small pieces with narrow mortises, you can clean out the waste with the saw, but on larger members I cut the waste out with a coping saw, after which I clean up the bottom of the mortise by chiseling to the gauged line.

Return the piece to the saw and lower the blade so that it cuts just to the bottom line marked on the end grain. Lowering the blade in tiny increments after each set of cuts will widen each mortise until the respective tenon will snugly slip through. Note the word "respective." Your markings for placement and position pay off here in eliminating a great deal of confusion (and probable error) by showing which mortise belongs with which tenon.

Theoretically, the final saw setting should produce mortises of identical width in each member, and each tenon should fit with the same snugness. Theoretically. In practice I've found that it's better to make the necessary adjustments to fit each joint separately. It takes only a little longer, and inspires more confidence in the strength of the finished assembly. A tenon offset with respect to its mortise or vice versa will not allow you to flip the piece over to make the second cut at the same saw setting. Rather, after cutting one kerf in each piece, you have to pause to make a new saw setting. Fine-tuning this second cut widens the mortise to fit its tenon.

At this point a skeptical reader might wonder how, if you can't expect the saw to cut precisely square to the fence, can you ever expect it to cut truly horizontally (in line with the workpiece)? Well, you can't. You just try to take advantage of this. Indeed, whenever I confront a situation where I doubt achieving the required accuracy, I try to determine on which side of perfect it behooves me to err, in order to obtain the best results. In this case, a snug joint and a frame that won't twist or wind will result if certain parts of the joint have a little looseness and other parts are correspondingly tight. Because there is some play in the horizontal indexing pin on my saw, I can manipulate the angle of error. By pushing up on the left side of the blade and then tightening the locking knob, I can minutely incline the blade. This will produce a tenon that will narrow ever so slightly from front to back, and a mortise that will widen ever so slightly from top to bottom. These "errors" make for a joint that looks, in a much exaggerated view, like that shown in the drawing at right. In actuality, the amount of deviation from horizontal amounts to less than half a degree (1/64 in. end to end), and the corner of the joint is tight and snug.

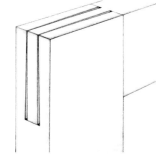

Before gluing up I always plane the rails clean on both sides. This removes the triangle markings, which you still need, so transfer them to the outside edges or lightly re-mark them immediately after planing. Also plane all inside edges on both rails and stiles. Apply the glue to both the mortise and the tenon, assemble the frame and draw it up first one direction and then the other with a bar clamp. If the shoulders were cut square and the bottom of the mortise pared true, and if the outside corner edge of the joint is tight, there is no need to leave the frame clamped up, especially since clamps can throw the frame out-of-square or in winding. Having checked for squareness (see page 95), you should now clamp the cheeks of each mortise with two C-clamps, using one at each "loose" area. Use pads to protect the work. When the glue is dry, plane the stiles flush with the rails.

The dovetail slip joint—Shown at the top of the facing page, this joint is well suited to being made on the radial-arm saw. The first step is to make a cradlelike fixture to hold the rail (tenoned member) at the proper inclination for cutting the tenon cheeks and shoulders. As shown in figure 3, the fixture is made from two pieces of 3/4-in. plywood and a couple of stretchers. The angle of inclination shown at a is the slope of the dovetail. This angle will vary, of course, depending on the dimensions and proportions of the stock. Two members joined with their edges in the same plane (figure 1) will not require as steep a slope as two members joined with their faces in the same plane (figure 2). A slope of 10° from the vertical will do in the former case, while a slope of 3° to 5° seems appropriate for the latter.

Marking out the pieces proceeds in basically the same manner as with the slip joint. Start with the rail and mark the shoulder lines.

The dovetail tenon—With a sliding bevel and a scriber, mark out the dovetail tenon, making it no less than 3/16 in. thick at the top edge. The sawblade will normally remove at least this

much from the narrow part of the mortise. Also remember that at the widest part of the mortise the cheeks are the narrowest, so leave enough wood here for a strong joint, something to consider when laying out the tenon. As a last step, witness-mark the outside edges of all the members.

Now place the fixture on the saw table, its high side against the fence on the left side of the blade. Place a rail, outside edge out (facing you), in the crotch of the fixture, and set the saw to the proper depth for cutting the shoulders (G). Clamp a stop block to the table for repeated cuts (H). To cut the opposite shoulders on the rails, you must reverse the fixture, placing the low side against the fence. Because in this position there is no back stop to hold the rail, you have to nail a strip behind the workpiece or devise some other means of holding it firmly in place. Don't try to make this cut without securing the work, or the blade will snatch the piece from your grasp and send it flying.

When you have cut all the shoulders for the tenons, tilt the saw to its horizontal position, and arrange the fence, spacer and auxiliary table in the same manner as for the slip joint. Set the fixture's high side against the fence, and place the workpiece face edge out in the crotch (I). Adjust the blade 1/32 in. above the top line scribed on the end grain for the tenon, and cut the tenon on one side; again advance the workpiece an inch or so at a time between passes, and stop the cut 1/4 in. shy of the shoulder. Break the waste piece off by hand and complete the cut. Cut the other side by reversing the jig and flipping the rail (J), taking the same cautions as before to secure the workpiece. Now, lower the blade 1/32 in. to the gauge line and trim the tenons to final thickness. To cut the shoulder at the bottom of the tenon, reset the saw and hold the piece directly on the auxiliary table.

The dovetail mortise—Gauge the depth of the mortise from the width of the tenon, and continue the mark across the entire edge of the stile (or leg). Notch the corner edge to determine the blade setting for the depth of the mortise. Lay out the shape of the pin on the face edge and continue the lines across the end-grain surface.

Set the sawblade now at the correct angle, using the sliding bevel (K). Unplug the saw before removing the guard; then replace the guard after making the setting. As in cutting the slip joint, you can tension the dovetail mortise and tenon by inclining the blade slightly, so that the tenons will narrow from front to back and the mortises widen from top to bottom. To get the correct setting—a fraction of a degree over the angle of the fixture—some trial and error is required. Use a test piece.

After fine-tuning the tilt angle, you must readjust the fence and maple spacer so that the teeth of the blade will just touch their edges. Line up the notch on the corner edge of the stile with the edge of the maple spacer, and clamp the stop block to the table. Adjust the blade to cut just above the bottom line marked on the end grain of the piece. Because the blade is set at an angle, you cannot cut the piece in increments, but must make the entire cut in one pass. This means that the blade will want to push the piece away, and so you must pull the saw into the cut very gradually, and with the utmost caution. When you have established kerfs on both sides (L), a coping saw removes the waste, and a chisel cleans the bottom of the mortise. Trim the mortise by lowering the saw in small amounts until the respective tenon slips snugly

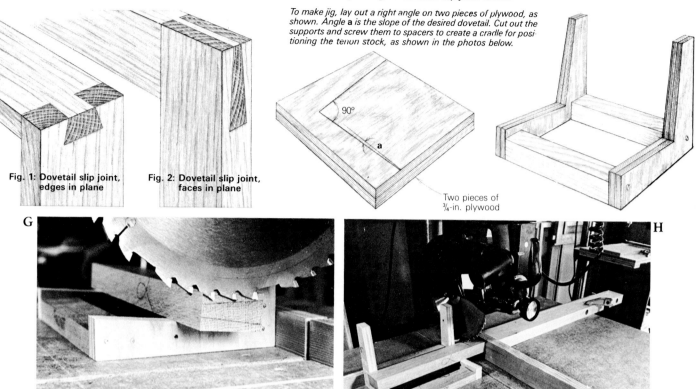

Fig. 3: Jig for dovetail-slip-joint tenons

To make jig, lay out a right angle on two pieces of plywood, as shown. Angle **a** is the slope of the desired dovetail. Cut out the supports and screw them to spacers to create a cradle for positioning the tenon stock, as shown in the photos below.

90°

a

Two pieces of ¾-in. plywood

Fig. 1: Dovetail slip joint, edges in plane

Fig. 2: Dovetail slip joint, faces in plane

G

H

To cut the tenon for a dovetail slip joint, author uses a cradlelike fixture, the construction for which is shown in figure 3, above. Place the scribed stock in the fixture and set the depth of cut for the shoulder (G). Then clamp a stop block to the saw table and cut the shoulders for one side of all your stock (H). To cut the shoulders on the other side, reverse the cradle so its low side is against the fence.

I

J

FRONT

BACK

When all the shoulders are cut, saw the cheeks on one side of each piece (I): tilt the saw to its horizontal position, set the fixture's high side against the fence and adjust the blade just above the scribed line on the end grain. To saw the cheeks on the other side, reverse the jig and flip the workpiece (J). Saw in a series of passes, advancing the workpiece between passes and breaking off the waste before completing the cut.

K

To saw the mortise for the dovetail slip-joint, the stock is positioned horizontally, and a sliding bevel is used to set the sawblade at the correct slope (K). Because the blade is tilted, the mortise cheeks must be sawn to full depth in one pass, necessitating extreme caution, lest the blade throw the work. The substitute fence, spacer and auxiliary table, used for cutting the simple slip joint, are used for this operation too. Flip the board to saw the other cheeks (L), then remove the waste with a coping saw and chisel.

L

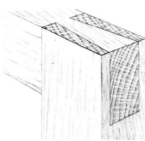

Variations on the slip-joint that the radial-arm saw can make include the dovetail slip-joint with two half-pins and a full tail (right) rather than with two half-tails and a pin, as described in the previous pages. Below is a dovetail three-way case joint for the four top corners of a frame-and-panel cabinet. The double dovetail pins on the front horizontal member are cut by hand.

Curtis Erpelding

in. If you have made correct settings, the joint will be tighter at the top corner than at the bottom. This tension actually forces the tenon into the mortise.

Other joints—In addition to the two joints described here, other variations are possible, either using part of the method or adapting it to different requirements. Instead of joining two members with a dovetail pin (tenon) and two half-tails, you could use the same fixture to cut a joint that consists of a full tail (tenon) on one member and two half-pins on the other, as shown in the drawing at left. In some situations such a joint might be structurally or visually preferable.

Sometimes it's not possible or even desirable to cut the entire joint on the saw, as was the case with the curved-stretcher table shown in the photos on page 39. Because the stretchers (rails) are curved, and glued together in a triangular fashion before the tenons are cut, I couldn't hold them on the saw table, certainly not in the fixture. I cut the mortises in the legs using the radial-arm saw, marked out the tenons from the mortises, clamped the stretcher assembly in my vise and cut the tenons with a backsaw.

I used a dovetail three-way case joint (photo, left) for the four top corners of a frame-and-panel cabinet. First I cut the dovetail slip joint in the regular way and then cut the two sockets in the mortised members, all on the radial-arm saw. From the two sockets I laid out the two dovetail pins on the third member, and cut these by hand. Next, I glued up the slip-joined members and cut the tenon flush with the socket walls to receive the pins.

The angled fixture pointed the way to an even more specialized joint—a knockdown, through-mortise-and-wedge joint. The fixture held the stretcher at the proper angle while the tilted saw cut a tapering dovetail mortise. A compound tapered wedge draws the stretcher tight against the cross-member of the legs. I applied this same joint to a sturdy knockdown leaning shelving unit (see pages 114 to 117), which uses minimal material and no metal hardware. □

Curtis Erpelding makes furniture in Seattle, Wash.

On exposed joinery

Architect Louis Kahn once said that the joint is the beginning of ornament. He was talking about architecture, of course, but the same can also be said about cabinetry and furniture. Until the Arts and Crafts Movement began in England, somewhere around 1860, the history of furniture design had been a history of hiding the joint. There are some notable exceptions to this generality. The American Shakers, whose religious scruples proscribed ornament, relied on visible joinery to give their simple designs character and presence. English country craftsmen, who inspired Ernest Gimson and others in the Arts and Crafts Movement, cared chiefly about the practical utility of their furniture, and had no reason to conceal its structural integrity beneath floral decoration and classical moldings. After Gimson and the Barnsleys, exposed joinery became, for better or worse, a design principle. Through wedged tenons and through dovetails attest both to the skill of the craftsman (if it's going to show, it had better be sweet) and to the honesty of the design.

The modern craftsman has the advantage of using power tools which greatly facilitate the speed and accuracy with which open joints can be made. And these tools do this, I think, without sacrificing any measure of handmade quality. Exposed joinery can be the signature of a craftsman or shop, an important design detail, and a record of the piece's manufacture. Industry can-

not economically use exposed joinery as a design element to any great extent. The few examples—machine-cut dovetails and finger joints in chair frames—usually lack the crispness, the clarity, and the careful proportioning that the individual craftsman can bring to a piece. When the joint fits really tight, when no gap shows and even the glue-line disappears, the end-grain and flat-grain surfaces set up a visual vibration, a dancing of surfaces. Conceptually there's magic in the tight geometrical mating of two elements, the triangular interweave of dovetails, the knotlike locking of mortise and tenon. Joinery can indeed be the beginning of ornament, but it can be the culmination of it as well.

—*C.E.*

Routing Mortises
A simple fixture and the right router

by Tage Frid

A mortising machine is an important piece of equipment, and in a cabinet shop with several workers one might be a good investment. Whatever kind you buy, a chain-saw or a hollow-chisel mortiser or a long-hole boring machine, you can expect to pay a lot for it, $2,000 or more. But by building a simple fixture for holding the stock to cut the mortises with a plunge router, you can have a mortising setup that works just as well as an expensive machine. The cost is only about $350, and you'll have acquired a heavy-duty router for general shop use.

Two makes of plunge-type routers are sold in the United States—The Stanley (models 90303 and 90105) and the Makita (model 3600 B). The Stanley plunge-base routers are production tools specifically designed for rough cut-out work, like cutting out holes in countertops for kitchen sinks or lavatories. They plunge to a set depth and lock automatically, but cannot be locked at any depth in between. Stanley, as you probably know, has sold its power tool division to Bosch Power Tool Corp. (PO Box 2217, New Bern, N.C. 28560), although some of the tools are still being sold under the Stanley label. The Makita router is very similar in design to the Elu router, which is a popular tool in Europe but isn't sold in this country

because it has a 220v, 50-Hz. motor, and would burn out if plugged into an American outlet.

Several things about the Makita model 3600 B make it a good router for mortising. It's a plunge-type router with a 2¾-HP motor. The body (motor/spindle assembly) is attached to a rectangular base by two ⅞-in. dia. steel posts. These fit into sleeves in the body, which can slide up and down on the posts (against spring tension), and be locked at any height. Plunge routing lets you begin and end a cut in the middle of a piece of stock without having to lower or lift the base from the work. With the motor running you can lower the bit into the wood by pushing down on the router's handles.

The Makita is pretty heavy (11 lb.) and well designed. The switch can be worked without having to move your hand from the handle, but it's not located right on the handle where you might turn it on accidentally when picking up the router. The body is locked on the posts by a latching lever instead of a knob, and you can reach this lever to raise or lower the bit without having to take your hand off the handle. An adjustable knob on the top stops the upward travel and also controls the depth of cut for ordinary routing.

For plunge routing there are two depth stops that let you

For cutting clean, precise mortises quickly, Frid uses a Makita plunge router with an easy-to-build fixture (below, left) for holding the stock and guiding the tool. Using appropriately sized or contoured supports, almost any piece of stock, like the curved chair back shown here, can be mortised in this fixture. At right, Frid tightens one end stop, which, along with the other, will limit the travel of the router and determine the length of the mortise.

remove stock in two passes rather than in one, and the stop rod is capable of both fast and fine adjustment. You can also control the depth of cut when plunge routing without using the stops. Just turn on the router, release the lock lever, push the bit down to the desired depth and lock it. The fence is secured with only one wing nut and is easy to set. The Makita router comes with ¼-in. and ⅜-in. adapter sleeves for its ½-in. collet, so you can use different bit sizes and cut mortises of almost any width and up to 2⅜ in. deep. By turning the stock over you can produce a through mortise up to 4¾ in. deep. The weight of the machine makes it stable while running and helps give you a smooth cut. You might find that the posts bind in the sleeves if you try plunging the router when the motor is not running. But when the motor is on, the vibration allows the body to move smoothly up and down on the posts.

The mortising fixture I made looks like a big miter box. Its length and depth can be varied to suit your needs, but for general use it should be 20 in. to 36 in. long and about 4 in. deep. This will let you mortise everything from chair legs to bedposts. Regardless of the length, the inside width of the fixture should not be more than 3¼ in., or else the router base will not rest on both side pieces. Make the bottom of the fixture from two pieces of ¾-in. plywood. It needs to be thick to give the sides of the box a large gluing surface and to hold them stiff with a minimum of flex. The edges of the bottom piece must be a true 90° to the face, or the sides will not be perpendicular and your mortises will be askew to the face of your stock. When gluing up, be sure the bottom edges of each side are flush with the bottom of the base. Locating pins will help hold the sides in alignment when tightening the clamps. Solid wood could warp, so you might want to make the sides from ¾-in. plywood with solid lipping on the upper

edges where the router base rubs. When the glue is dry, check the two upper edges with a square; they must be perfectly parallel. If they are not, take light passes on the jointer until the edges are square and parallel, or use a hand plane to do the same thing.

You need to install lateral end stops on the top edge of the inboard side of the fixture. The two stops are slotted strips of wood with shallow grooves on their bottoms to fit over and ride along the edge. The slot in each strip rides around a ³⁄₁₆-in. stove bolt which engages a T-nut embedded in the side. A barrel nut would prove even more durable and easier to install. Also, you could drill pilot holes in the upper edge of the inboard side for hanger bolts; use wing nuts with these to tighten down on the stops.

To cut the mortises in a piece of regular dimensions—a straight table leg, for example—raise the workpiece in the fixture so it's almost flush with the top edge. Center the area to be mortised in the middle of the fixture and clamp it to the inboard side. Set the stops on both ends to contact the router base so the bit can travel the full length of the mortise in one pass. With the fixture held in a vise, set the fence on the router the right distance from the bit and butt the base against the left-hand stop. Switch on the power, release the lock lever and lower the bit into the wood. For a ⅜-in. bit, a ¼-in. depth of cut would be safe. Then pull the router to the right. Don't start at the right-hand stop and push the router to the left. By pulling the router left to right, the rotation of the bit will hold the fence against the side of the fixture, which will give you a good, straight cut.

During routing, dust and shavings can get compacted on the ends of the stops where the router base makes contact. If you don't keep this debris cleaned off, your mortise will get

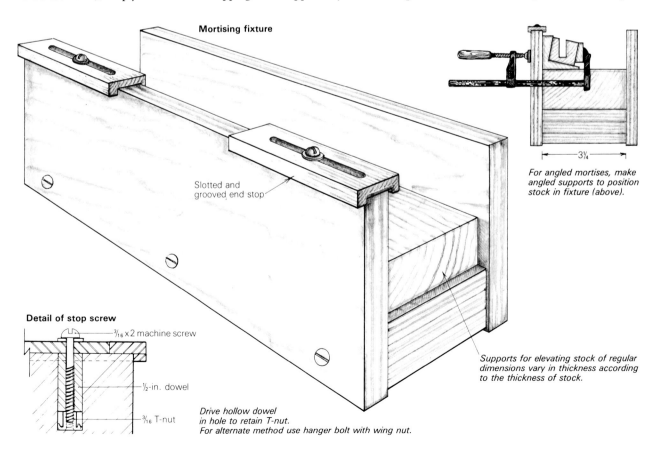

Mortising fixture

Slotted and grooved end stop

Detail of stop screw

³⁄₁₆ x 2 machine screw

½-in. dowel

³⁄₁₆ T-nut

Drive hollow dowel in hole to retain T-nut.
For alternate method use hanger bolt with wing nut.

For angled mortises, make angled supports to position stock in fixture (above).

Supports for elevating stock of regular dimensions vary in thickness according to the thickness of stock.

3¼

The fixture shown here was made especially long for mortising bed-posts. The stock is held in place by a wedge at either end, and the lateral end stops are set to limit the length of the mortise. By making a full plunge cut at the extreme ends of the mortise and routing out the waste between with a series of shallow passes, mortising proceeds with speed and precision.

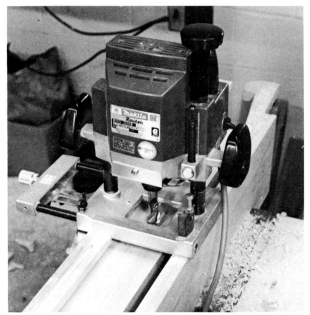

Equipped with a spiral end mill, the Makita model 3600B router is an excellent tool for mortising. But its powerful motor and square base make it also well suited for clamping upside down in the tail vise of your bench, where with the proper fence it becomes a spindle shaper.

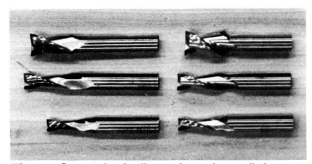

These two-flute spiral end mills were designed especially for routing wood. Bits with cutting diameters of ½ in. or more can cut as deep as the bit will plunge, but bits with cutting diameters smaller than ½ in. are limited in their depths of cut by the diameter of the shank. The bits on the right in ⅜-in., ½-in. and ¾-in. cutting diameters are made by Onsrud Cutter Mfg. Co. The bits on the left, in corresponding cutting diameters, are made by Ekstrom Carlson Co. Their longer shanks give them greater depth-of-cut capability.

shorter as the cut gets deeper. One way to avoid the problem is to make an initial plunge cut to full depth with the router held against the left stop and then against the right stop. Then you can rout out the waste between in several passes without having debris pack up against the stops.

The depth of each cut depends on the hardness of the wood you are cutting and on the size and kind of bit you use. Make repeated cuts, always left to right, lowering the bit between passes, until you have reached the desired depth for your mortise. All this might sound complicated, but you will be surprised at how fast it works. I have found it faster and cleaner cutting than the hollow-chisel mortiser.

To cut angled mortises in regular stock, like those in chair legs to receive tenons on stretchers and rails, make angled supports to hold the stock in the correct relation to the bit. To mortise curved pieces—a chair back, for instance—bandsaw a piece to fit the side of the curve opposite the cut and use it to support the stock when clamped in the fixture. You can place the curved support under the stock for mortising on one side. To mortise the adjoining face, support the stock from the inboard side of the fixture using the same curved piece, and a flat support on bottom.

For general mortising, I use two-flute, straight-face bits with ¼-in. shanks. High-speed steel, straight-face bits will work, but they will get dull faster than carbide bits. For mortises ½ in. wide or more, you can use bits with ½-in. shanks, which will perform better than bits with smaller shanks because they are stiffer and will chatter less. The best bits for cutting large mortises are two-flute spiral end mills with ½-in. shanks. They are especially designed for plunge cutting and for fast chip removal, and because of their spiral form they have a shear-cutting action. When wasting the area between the two plunge cuts, spiral bits can make passing cuts as deep as ⅜ in. without protesting. But they will start to scream when you make a passing cut that's too deep, and you will find yourself forcing the bit into the work. This is not good. Spiral end mills will cut effortlessly in a straight plunge and when taking a lateral pass that's not too deep.

You can get spiral end mills made for routing wood from Onsrud Cutter Mfg. Co., 800 E. Broadway, PO Box 550, Libertyville, Ill. 60048 and from Ekstrom Carlson & Co., 1400 Railroad Ave., Rockford, Ill. 61110. Costs vary, but generally ½-in. dia. and ⅜-in. dia. bits are under $10, while ¾-in. dia. bits run about $20 (1981). □

NOTE: The business of setting the router stops and of locating stock in the fixture can become tedious when cutting a lot of mortises. Here is a solution: First, knife a vertical line on the inside face of the fixture, near its center. This is the primary reference for subsequent measurements. Next, scribe a stop line on the top edge of the fixture, to the left of the centerline, the precise distance from the cutting edge of your mortising bit to the edge of the router base. With the left-hand stop locked at the stop line, a knife line on the stock marking the left end of the proposed mortise can be brought to the vertical centerline. Now, with the router placed against the left-hand stop, measure over on the fixture's edge, from the right side of the router base, the length of the proposed mortise minus the bit diameter. This locates the right-hand stop. To set the depth of cut, lower the adjusting knob until the cutter grazes the stock surface. Then set the depth screw to the depth of the mortise above its stop. Finally, back off on the adjusting knob so the bit will clear the stock. To make all of these measurements quick and reproducible, you can mill a set of hardwood gauge blocks. Instead of measuring, you simply insert the correct gauge block between the router base and end stop, and between the depth screw and depth stop. In addition to cutting mortises, these gauge blocks will come in handy for other setups in the shop.

The Frame and Panel
Ancient system still offers infinite possibilities

by Ian J. Kirby

Scarcely another system in the whole range of woodworking has more variation and broader application than the frame and panel. In the frame-and-panel system, pieces of solid wood are joined together into a structure whose overall dimensions do not change. The frame is usually rectangular, mortised and tenoned together, with a groove cut into its inside edge. The panel fits into this groove: tightly on its ends since wood does not move much in length, but with room to spare on the sides because wood moves most in width (figure 1). Wood is not uniform and as it moves in response to changing moisture conditions, it cups, twists, springs and bows. Trapping the panel in the groove inhibits this misbehavior.

Historically, the basic technique that made possible the frame and panel is the mortise-and-tenon joint, on which I have already written extensively. The frame and panel is a basic unit of structure. It can be used singly (a cabinet door) or in combination (to make walls, entry doors, cabinets). The several elements of a single frame and panel may be varied almost without limit, and its aesthetic possibilities are infinite. The little choices made during its manufacture are aesthetic choices, and we can begin to see the interdependence of design and technique. Neither the frame-and-panel system nor its joinery can be thought of as an end in itself. Neither has any importance except in application, toward the end of making a whole thing out of wood.

To overcome the panel's tendency to distort, we make it as thin as the job will allow, while we make the frame relatively thick. By doing this we haven't significantly altered the amount by which the panel will shrink and expand, but we have rendered it weaker so that the frame has a better chance of holding it flat. Panels can be made as thin as $3/16$ in. but such panels were uncommon before the 19th century because woodworking tools were not readily capable of such refinement. The thickness of the panel was dealt with in a number of ways, the most usual being to raise and field it.

Generally raising and fielding are thought to be the same thing—the process of cutting a shoulder and a bevel on the edges of the panel and thereby elevating the field—that is, the central surface of the panel. I'd like to distinguish between these two terms. Fielding refers to any method of delineating the field of the panel from the frame; raising means cutting a vertical shoulder around the field, which may or may not be accompanied by a bevel.

Given this system for maintaining panel size and shape against the hygroscopic movement of wood, woodworkers in the past found that the system's elements could be varied to produce different aesthetic results in terms of form, color, texture, pattern value, proportion of parts, highlight and shadow (figure 2). By carving the panel, we imprint upon it richness and grandeur. By inlaying it, we make it into a vehicle for the decorative use of other materials, and by using moldings of different profiles, we change the various propor-

Fig. 1: Section through frame and panel

Stile

Rail

Fig. 2: Proportion and pattern

The pattern possibilities are infinite. The peg holding the panel on center can be an interesting detail for the observer to discover—don't let it become a cliche. It needn't be taken through to the front of the rails.

Illustrations: Ian J. Kirby

tions within the whole. We can alter the mood of a piece by changing the panel's attachment to its frame.

Within the predetermined dimensions of the frame, the relative proportions of the rails and the stiles can vary considerably, though it is common to find the top rails in a door to be about two-thirds the width of the bottom rail, and the stiles to be about three-fourths the width of the top rail. This isn't necessary to produce an acceptable appearance, but many woodworkers build doors as though these proportions were divinely decreed. Altering them within the limits of structural necessity can produce a wholesome diversity of appearances. The proportions of the panel—the dimensions of its field, the depth of its shoulder, the width and slope of its bevel—are all subject to considered alteration.

For showing off highly figured wood, the frame and panel is excellent. The usual way of doing this has been to resaw the figured piece and to edge-join the halves into a bookmatched panel. Examples of discriminating and sensitive use of various frame-and-panel constructions can be found, for example, in the work of Edward Barnsley, England's grand old man of furniture designers. His work, as well as pieces designed by such members of the Arts and Crafts Movement as Ernest Gimson and his Dutch foreman, Peter Waals, illustrates the fact that generally the plainer form of paneling—overlaying it onto the frame—shows a highly figured piece of wood to its greatest advantage. Raising and fielding can look fussy when the wood is highly figured.

The face of the panel can be treated in countless other ways. You can inlay it with mother-of-pearl, ivory, brass or other materials, or the field can serve as a ground for marquetry or wood inlay. Carving on panels has taken many forms. English "joyners" of the 13th to 15th centuries imitated almost slavishly the tracery and linenfold patterns common in stonemasonry. It was also common in this period for panels to be painted with vibrant colors in a variety of abstract geometric designs. In our own century, Mousey Thompson adzed the surface of quartersawn oak panels for a mild rippled texture, subtle to touch and sight.

I wish contemporary panels conveyed such vitality and served the imagination as well. I'm not saying that we have to paint panels or inlay them or dress them up in other ways. I am saying that we ought to realize that the panel is an unexploited vehicle for expression, for there is in fact no woodworking system with as great a potential for individualization within its essential structural features. This invites the woodworker to explore some of the system's possibilities. The easiest way to play with all of these possibilities is to make a

full-size drawing, or several of them, to help you visualize the critical relationships between the parts. A further survey of traditional treatments and manufacture of the solid-wood frame and panel may help contemporary woodworkers to a livelier use of the system.

The frame — Designing a frame and panel begins with a sectional view of the two parts and a decision about the joint to be used in bringing the frame together. The groove is generally placed in the center of the frame stock. If you use a square haunched mortise and tenon to join the frame, then the grooves can be plowed right through the joint without interruption (figure 3). This is no happy accident. The joint was designed for the groove to be continuous, because a plow plane can't cut a stopped groove. When making the haunched mortise and tenon by hand, the question arises about what to do first, plow the grooves or cut the joint. I recommend cutting the mortises and tenons first. If you plow the groove first, you obliterate the gauge lines on the mortise and on the inner edge of the tenon. And nothing is more agonizing than trying to cut an accurate mortise without reference to gauge lines. Also, it's easy to be tricked into thinking that you are cutting the mouth of the mortise at the same time you plow the groove. You can in fact. But if your mortising chisel is not exactly the same width as the groove, you'll find it almost impossible to cut the rest of the mortise in an accurate way. Using a chisel that is too large or too small, even by a minute amount, makes cutting the mortise walls an exercise in guesswork and error.

If you wish to use a mortise and tenon of the sloping haunch type so that the joint shows an uninterrupted straight line, then some means of stopping the groove must be employed. This is quite easily done on a router, shaper or circular saw by the use of end stops.

Another aspect of the frame that deserves attention is the profile of its inner edges. Since standard shaper knives are frequently used to make molding and scribe cuts in rails and stiles, I'd like to describe the less common practice of chamfering the inner edges. It creates a boundary that is decisive and bold, not busy or blurry like poorly conceived moldings. There are three conventional ways to chamfer the inner edges of the frame, two of which require redesigning the joints.

The simplest way is to assemble the frame dry and cut a 45° chamfer with a router, using a bit with a ball-bearing pilot. The cut will be rounded in the corners (figure 4) but can be squared up if desired with a sharp chisel to form what is called a mason's miter (figure 5). But if you want the line of the joint and the outer edges of the chamfer to form one continuous line, then more refined joinery is called for. The first method involves cutting the tenons on the rails as if you were making a long-and-short-shouldered joint (figures 6 and 7). Cut the mortise in the conventional fashion, and then chamfer the inner edge of the stile. To accommodate the chamfer, the long shoulder of the tenon is beveled inward from the shoulder line. A variation of this method also leaves the line of the joint and the outer edge of the chamfer uninterrupted. But instead of beveling the underside of the long shoulder on the rail, you remove a section from the face of the stile to accommodate the long shoulder. Then the chamfers on both rail and stile must be mitered to fit (figure 8). Making this joint requires careful measuring and marking with a mortise gauge and cutting gauge. Chiseling should be done across the

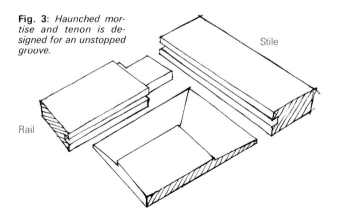

Fig. 3: *Haunched mortise and tenon is designed for an unstopped groove.*

Stile

Rail

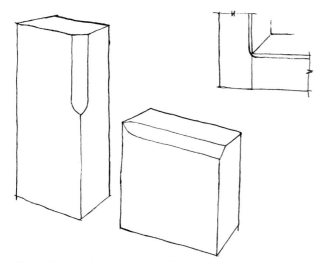

Fig. 4: *To bring the chamfer around the corner in a 90° arc, cut most of the bevel before assembly. The corner round should be completed after gluing up because of the fragile short grain on the rail.*

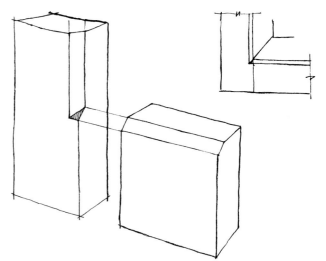

Fig. 5: *Mason's miter brings vertical and horizontal chamfers together. The disadvantages are that the chamfer does not line up with the shoulder of the joint, and a little triangle of end grain spotlights the mitered corner.*

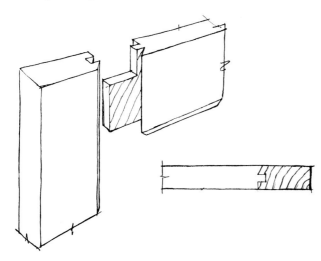

Fig. 6: *Beveled or scribed version of long-and-short-shouldered mortise and tenon permits precise alignment of joint and chamfer lines.*

Fig. 7: *Plan of stile/rail joint showing original line of shoulder before undercutting.*

Scribing line

Original shoulder line

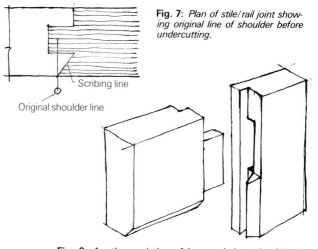

Fig. 8: *Another variation of long-and-short-shouldered mortise and tenon demands care in marking out to get the joint to close properly and the miters to meet on a line.*

grain, and the mouth of the mortise plugged with a softwood block to prevent the tissue from splintering out.

Don't sand the chamfer if you wish to retain its crisp edges and the fine texture of its tooled land. Sandpaper rounds and softens these critical light-reflecting angles and faces. You may try sanding, and even get an apparently satisfactory look until the finish is applied, but then you will find the crispness lost and the clarity of the land muddied.

Yet another variation in the design of the frame is not to use grooves at all, but to lay the panel in a rabbet cut into the edges of rails and stiles. The panel is retained with an applied molding on the show side of the work. Some might think this method to be a good example of bad workmanship, but it is well suited for some types of work, including those using modern materials. It's worth noting that much of the traditional frame-and-panel joinery was done this way.

The panel — When thinking about the design of a panel three possibilities are available: a fielded panel, raised or not, an unfielded panel and an overlapped panel. According to the type of design you choose, you must consider variables

such as the size of the field relative to the frame dimensions, the width and slope of the bevel, the depth of the shoulder, and the treatment of the field (carving, inlay, figured wood) — all of which combine to determine the finished look.

You can field a panel without raising it by cutting the bevel right through to the field and eliminating the shoulder. But it's difficult to get good results when you define the field with the bevel alone because invariably that critical line where the two converge will be untrue. It can be straightened out with a hand plane, but it's not easy. Also, panels without shoulders look indecisive, even if the field edge is straight.

It's common practice when raising a panel to cut the shoulder lines first (figure 9). This fields and raises the panel and gives you an idea of its proportions. You can accomplish this by setting the fence on your bench saw to the width of the bevel and the blade to the depth of the shoulder, producing four shallow cuts equidistant from the panel's edges. Saw the bevel by tilting the sawblade to the required angle and setting it to the required depth. When cutting the bevel, a wedge should fall off the waste side of the cut; this means that pressures on the blade are equal right and left. But raised

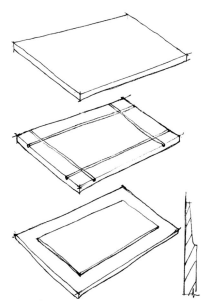

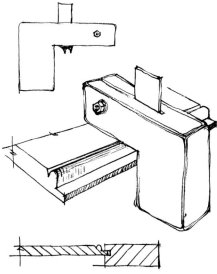

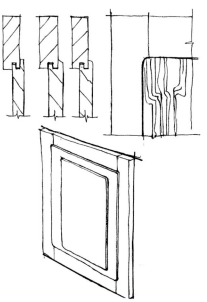

Fig. 9: *The field is defined by saw or router cuts. You can use a nosed cutter in your router to give the shoulder a softer profile. The bevel is angled at the time of sawing or it can be cut with a bench rabbet plane.*

Fig. 10: *A panel edge need not be beveled. Here it is rabbeted and a bead mold run down its edge. The scratch-stock must be held firmly and pushed away from the body. It should settle gently into the cut and not be forced.*

Fig. 11: *The edges of the overlapped panel may be treated in a variety of ways—rounded, beveled or left square. At this stage you introduce details that affect the highlights and shadows of the panel.*

panels are usually more delicate than this will allow. So to minimize blade vibration and consequent scoring of the bevel, use a sharp, stiff tungsten-carbide-tipped saw and feed the work gently into the blade.

However carefully you cut the bevel, its sawn face will need some cleaning up. It can be sanded, but you'll have greater control and get better results using a rabbet plane that's wide enough to clean the whole width of the bevel in a pass. You could also make a jig for your router and make the cuts with a properly profiled bit, or you can obtain the fastest results with a spindle shaper. The latter should be used with great caution, and you should make three or four passes, removing only a little stock with each.

Because the whole system is designed to accommodate the hygroscopic movement of wood, it's not unusual for a panel to travel about in its grooves, sometimes being noticeably off center. This is easily rectified by driving two nails or wooden pegs into holes on the inner edge of the frame, top and bottom, so that they capture the panel on center (figure 2, page 48). The pegs make interesting little details if left slightly proud and rounded off. You can get the same result by applying a dab of glue in the center of the grooves top and bottom, taking care during assembly to position the panel properly. It is normal for the panel to rattle when tapped. The rattle can be eliminated, but need not be.

The panel doesn't have to be raised and fielded to be held in the frame. One of the sweetest systems brings the groove forward of center to create a flush panel from very thin stock. This system demands accuracy in cutting the top and bottom shoulders so they just touch the rails. The panel edges, which move slightly over the frame, can be molded by using a scratch-stock that you can make yourself (figure 10).

The third possibility exchanges the angular, vigorous look of the usual raised panel for the softer, more subdued look of the overlapped panel. Panels of this sort are rarely seen because most woodworkers hesitate to depart from the more accepted method, yet they are no more difficult to make. The panel is held in the frame by a set of tongues and grooves, which should not be cut too deep as a long tongue is likely to curl back (figure 11).

Given the possibilities of the frame and panel, it is surprising to find them so little realized by contemporary cabinetmakers working in solid wood. Industry, however, has not ignored their appeal. In one method of quantity production, frame-and-panel doors are molded from a mulch of the sort used to make particle board. After a few seconds of heat and pressure, out pops a frame and panel, raised, fielded and detailed to your requirements, ready for printing with a photocopy of wood grain. Before the offended reach for their pens, consider that if industry will apply its technical and economic resources to such an extent, there must be a strong demand for the frame and panel. Given this market, there is a noticeable lack of frame and panel being used in a refined and exciting way by makers of hand-built, solid-wood furniture. This seems a pity since I've always felt that the cabinetmaker's shop could be the birthplace of technical and aesthetic models for industrial production.

The sad fact of the matter is that there exists an emotional antagonism between the designer/craftsman and the designer/executive, and little productive communication is shared between them. The craftsman suffers the most as a result, because he closes the door on salespeople, designers, decorators and others who work either with or for the larger furniture manufacturers. It does the craftsman little good to have a vast technical knowledge, a keen aesthetic sense and a shop full of tools if he denies himself contact with people who can market his furniture and who can benefit from his fidelity to quality and his innovative thinking. I hope this examination of the frame and panel will promote improved communication between the craftsman in his shop and the larger world of woodworking. □

Radial-Arm Raised Panels
You can even make them out of plywood

by William D. Lego

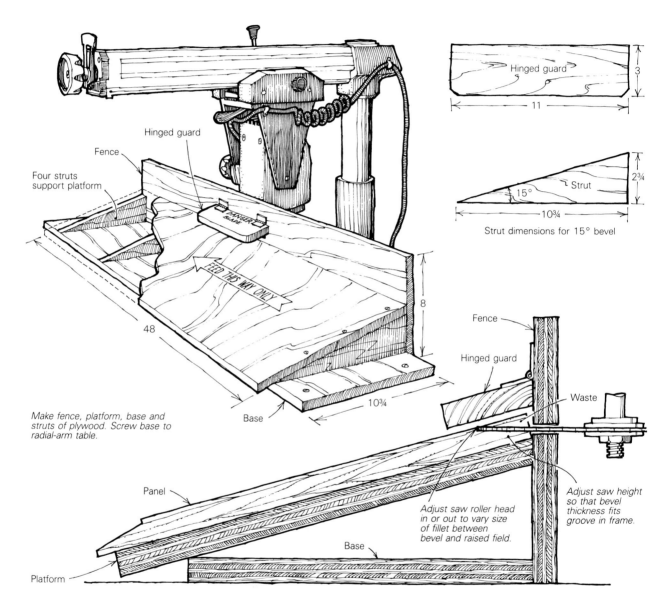

Hinged guard

3

11

15° Strut

2¾

10¾

Strut dimensions for 15° bevel

Fence

Four struts
support platform

Hinged guard

Fence

Hinged guard

Waste

*Make fence, platform, base and
struts of plywood. Screw base to
radial-arm table.*

48

8

10¾

Base

Panel

*Adjust saw roller head
in or out to vary size
of fillet between
bevel and raised field.*

*Adjust saw height
so that bevel
thickness fits
groove in frame.*

Base

Platform

I use raised panels in much of the cabinetry I build, but like many small-shop woodworkers, I can't justify the expense of buying a large shaper just for this purpose. I probably couldn't shoehorn one into my shop anyway. Instead, I designed this jig that allows me to cut all sizes of raised panels with my radial-arm saw. As shop aids go, this one is practically bullet-proof—you can make it out of scrap, set it up in no time and, when it's not in use, hang it up on the shop wall, out of the way.

I built this jig two years ago, and I've found that it has two advantages over a shaper: it's safer to operate and the panels have smoother, splinter-free bevels. This last point is important to me because I make my raised panels out of hardwood

plywood and then cover the exposed edges of the plies with veneer backed by a thermosetting adhesive. This veneer, which I buy in rolls and sheets from Allied International Inc., PO Box 56, Charlestown, Mass. 02129, can be applied with a hot iron. The technique may not delight the purist, but with all the crooked, twisted lumber we seem to get these days, using plywood saves the time and frustration of gluing up solid stock and then milling it flat. This method seems best for panels that will be painted, but if you apply the veneer carefully, you can get decent results with clear finishes. Of course, the jig works just as well with solid wood panels.

As the drawing shows, my jig consists of an inclined platform mounted on a base that can then be screwed or clamped

Drawings: Jim Richey

A 2-in. PVC elbow, left, mounted behind the fence and connected to a shop vacuum collects the dust from Lego's panel-raising jig. He glued the elbow in place before slowly sawing its blade slot. Fence cutout accommodates saw motor when sanding bevels. The photo below illustrates safe hand position for feeding a plywood panel past the sawblade. Hinges attach the guard to the fence, allowing it to pivot up slightly, so offcuts won't jam.

Exposed plies on the bevels of a plywood panel won't do, but once they are covered with adhesive-backed veneer tape, above, they look fine and they will take an excellent paint finish. The veneer tape is applied with a hot iron. Where it meets at the corners, Lego razors a neat miter joint, below.

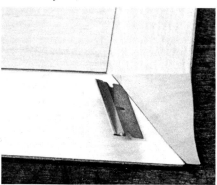

to the radial-arm table. You can experiment with the platform angle, but 15° is a good starting place. I'm kind of a safety nut, so I installed a hinged guard above the point where the blade projects through a slot in the jig's fence. The guard has to pivot only slightly, but don't use a rigid guard—the panel offcuts might jam between the blade and the guard. To collect dust and chips, I used construction adhesive and a few brads to mount a 2-in. PVC elbow in the fence behind the blade, then connected the elbow to my shop vacuum, as in the photo at top left.

Once you've built and positioned the jig, make some test cuts and adjust the height of the saw until the panel bevel tapers to a perfect fit in the grooves of the frame it will go into, as shown in the drawing at left. To vary the size of the fillet between the bevel and the panel's raised field, move the saw roller head in or out. Steel combination blades work well for panel-raising, but because they dull quickly, I find that I have to install a sharp blade after raising a half-dozen panels or so. A carbide-tipped blade would last longer. If I'm using solid wood, I sand the sawmarks off the bevels using a Sears 8-in. sanding disc with a 2° taper (catalog number 9 GT 2274). This tapered disc contacts the wood with a small, conical-shaped section instead of the wide, swirling arc of a flat sanding disc, thus leaving fewer sanding marks. You don't have to sand the bevels of plywood panels, since they'll be covered by veneer tape anyway. □

Bill Lego owns and operates a six-man cabinet shop in Springfield, Va.

Decorative joint enhances frame

For my cabinet doors I use a simple frame-and-panel construction with mortise-and-tenon joints. The method described here allows me to add a decorative molded profile to the inside edge of the frame, without having to cope the molding joint at the corners or to go to the trouble of making a masons' miter.

First I mill all my stock to the same thickness, cut the frame members, mortise the stiles and tenon the rails. After the joints have been fitted, I plane $1/16$ in. or so off the face of each rail. This step puts the face of the rail on a different plane than the face of the stile. Now you can rout a shallow (no deeper than $1/16$ in.) profile on the inside edges of both stiles and rails. Of course, the decorative molding runs the length of the stiles rather than stopping at the inside edge of the frame. The effect, though not traditional, is handsome, and the resulting doors are strong and light.

—*Pat Warner, Escondido, Calif.*

The Scribed Joint
Masking wood movement in molded frames

by Morris J. Sheppard

When moldings meet at an inside corner, as in framing a paneled door, they can be mitered by cutting each piece to a 45° angle. The joint is quick and it looks fine...until wood movement inevitably opens up the miter. Scribing the joint is an alternative to the miter. In this method, you cut the rail molding to the exact reverse section of the molding it will overlap on the stile. This allows the wood to move without breaking the joint. Where a center rail meets a stile the scribed moldings will slide and remain tight even with seasonal movement. In this article, I've used an ovolo molding or "sticking" on the frame parts, as shown in figure 1. However, any molding except those with undercuts can be scribed. Undercut molding must be mitered. A version of the scribe called the cope-and-stick joint can be done on the shaper or tenoner, but you can get excellent results by scribing with hand tools.

You'll need a small backsaw, chisels, a gouge and a miter template. A commercially-made template is brass, about 5 in. long and cut to a 45° angle at each end. You can make your own out of wood, but make sure it is dead accurate. Ovolo sticking is scribed with an in-cannel gouge whose radius matches that of the molding. You'll need a gouge to match each size of molding you want to scribe. Put a keen edge on the in-cannel by working the inside bevel with a slip stone and then remove the burr by holding the outside of the tool flat against a benchstone. Don't double-bevel the edge as you would a carving gouge. If you do, it won't cut straight.

Prepare your framework as you usually do. In my shop, we mortise the rails and the stiles on the slot mortiser, then insert a loose tenon. You can use a dowel joint or a conventional mortise and tenon, but when cutting the rails remember that their shoulders fit to the bottom of the panel rabbet and not to the inside edge of the molding. Mill the molding and the rabbet along the full length of the rails and stiles.

Begin the scribe by cutting away the molding on the stiles where the rails meet them—at the stile ends and the center of the rail if a middle stile is used. The molding should be cut

The scribe joint's overlapping moldings hide wood movement.

back even with the depth of the rabbet and ought to align with the listel on the rail. To get an accurate mark, hold the rail against the stile and strike a knife line, as in figure 2. Then saw down with the backsaw, being careful not to go deeper than the rabbet's depth. Remove the waste by paring with a chisel or by bandsawing.

Now move to the rails. Place the miter template over the molding at the end of the rail as in figure 3. Align it so the miter cut will end exactly at the tenon shoulder; use the rail listel as a guide. Fix the template with a clamp and then use a sharp chisel to pare away the waste. On the final cut, rest the back of the chisel firmly on the miter template. With the miter cut, the contoured edge of the molding outlines the scribe cuts, which are then made perpendicular to the edge of the rail. Cut straight down with a straight chisel at the listel, and with the in-cannel gouge make the concave shape that will mate over the stile molding (figure 4). Several cuts may be needed but the trick is to make the last cut precisely at the mark outlined by the miter. It helps to stand directly above the work with light from the side, casting a shadow at the outlined edge. If the gouge is keen, it will be easy to place it right on the line, and a firm push will be all that's needed. Be careful with the thinnest corner of the scribed molding, as it is prone to damage. Hold the gouge square to the work or a gap will show in the finished joint. Use a small chisel to clean up the bottom of the cut.

The scribe joint will also work in frames with a groove for the panel instead of a rabbet. Then the molding on the ends of the stiles gets cut away to the bottom of the groove and one tenon shoulder is offset to accommodate the rear wall of the groove (*FWW #18*, p. 88).

Cut accurately, the scribed pieces should slide together perfectly, as in figure 5. And they should stay that way through many seasons of wood movement. □

Morris J. Sheppard designs and makes furniture and cabinets in Los Angeles, Calif. Photo by the author.

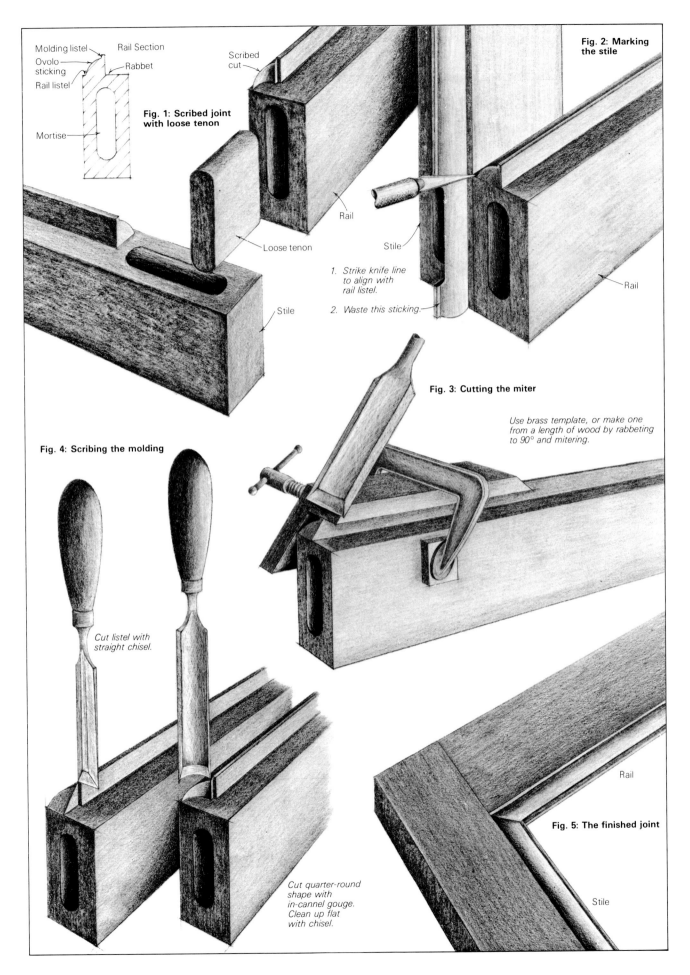

Molding listel
Ovolo
sticking
Rail listel

Rail Section

Rabbet

Mortise

**Fig. 1: Scribed joint
with loose tenon**

Scribed
cut

Rail

Loose tenon

Stile

Stile

**Fig. 2: Marking
the stile**

1. Strike knife line
to align with
rail listel.

2. Waste this sticking.

Rail

Fig. 3: Cutting the miter

Use brass template, or make one
from a length of wood by rabbeting
to 90° and mitering.

Fig. 4: Scribing the molding

Cut listel with
straight chisel.

Cut quarter-round
shape with
in-cannel gouge.
Clean up flat
with chisel.

Rail

Fig. 5: The finished joint

Stile

Entry Doors

Frame-and-panel construction is sturdy, handsome

by Ben Davies

Exterior doors are the problem child of architectural design. They are required to perform three functions: seal off an opening from the exterior air, open to allow passage and then reseal, and be attractive. All this from wood, a material that can change in size as much as an inch over the width of a typical opening. While each of these functions might be separately accomplished with ease, their combination into one design creates problems.

Single-panel board-and-batten constructions of edge-glued lumber are generally too unstable for exterior doors. They cast or wind unless great care is taken in the selection and seasoning of the lumber. They also expand and contract so much with the seasons that sealing against the weather is impossible. These shortcomings can be overcome by using frame-and-panel construction and, in fact, most doors are made this way. The style is relatively stable and offers great flexibility of design. Even the familiar commercial veneered doors are a variation of the frame and panel—the panel is reduced in

thickness to veneer and glued over the frame rather than inserted into grooves, and cardboard honeycomb or wood cores support the veneer. These doors succeed admirably in the first two functions a door must perform, but fail miserably at being attractive.

The frame-and-panel door has been in use for so long that its construction is well understood, and variations on its designs have been thoroughly explored. When any construction method remains dominant for hundreds of years it can mean only that it works quite well.

The standard size for entrance doors of new construction in the United States is 3 ft. wide by 6 ft. 8 in. high. A walnut door of this size can weigh 80 lb. to 100 lb. or more, depending on the thickness of the panels and the amount of glass. This is considerably heavier than a softwood or hollow-core door of the same size, and great care must be taken to ensure that the joints are well designed and well constructed.

I have seen a number of doors fail that were constructed

Routed sticking stops short of corners in raised-panel door, left, but continues around corners of flat-panel door, center. Both are Honduras mahogany. Detail from oak door, above, shows routed sticking combined with molding.

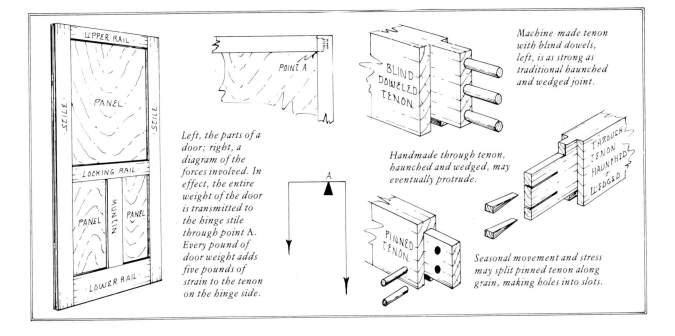

Left, the parts of a door; right, a diagram of the forces involved. In effect, the entire weight of the door is transmitted to the hinge stile through point A. Every pound of door weight adds five pounds of strain to the tenon on the hinge side.

Machine-made tenon with blind dowels, left, is as strong as traditional haunched and wedged joint.

Handmade through tenon, haunched and wedged, may eventually protrude.

Seasonal movement and stress may split pinned tenon along grain, making holes into slots.

with a mortise-and-tenon joint pinned through the cheek with dowels. A stronger joint is one with blind dowels inserted into the end of the tenon and bottom of the mortise. I use a 3-in. deep mortise and tenon with three or more 1/2-in. diameter blind dowels to join the stiles with the rails. Interior parts of the frame, such as muntins, are joined to the rails with smaller tenons, usually made to fit the groove cut for the panels, and are also blind-doweled. For flat panels I ordinarily use a 1/2-in. deep groove in the rails and a 3/4-in. deep groove in the stiles. I have found that this difference is usually sufficient to compensate for the greater shrinkage that occurs across the grain of the panels.

Several factors make blind dowel pins preferable to through-the-cheek pins. The first is visual. Dowels through the cheek are often chosen because they give the same sense of rigidity to frames as dovetails give to casework. While they do make a door look sturdy, the time will surely come when that particular effect is not wanted.

More importantly, I believe the blind-doweled tenon to be stronger than one pinned through the cheek. A tenon with blind dowels need not be haunched because the dowel pins not only make the tenon effectively longer, but also transform a stub tenon into a haunched tenon. Thus the glue area of the tenon becomes about one-third greater. And a stub tenon can be made more quickly than a haunched tenon.

Dowels perform two main functions. One is to prevent the tenon from sliding out of the mortise and the other is to counteract the bending moment of the weight of the door about the point where the tenon enters the stile. The lever arm through which the through-the-cheek dowel must act is necessarily about 3/4 in. shorter than that of the blind dowels. In a 3-in. tenon this difference translates into 25% greater strain on the pins. It is very important to understand that on a 36-in. door with stiles 6 in. wide, every pound of door weight adds about 5 lb. of strain to the dowels on the hinge side of the door. The wider the door and the narrower the stiles, the more intense the leverage. The maximum length of a through-the-cheek dowel is the thickness of the door, while the blind dowel can be twice as long as the width

of the stile minus the tenon length. The extra dowel length is significant because part of the glue line between the dowels and their holes is end grain joined to long grain.

I have not discussed the through wedged tenon because this joint must be made by hand, a relatively time-consuming operation. However, the joint is strong, although in the long run the tenon will protrude slightly from the stile.

The total strength of a blind doweled mortise-and-tenon depends on two factors: the shear strength of the glue line that joins the cheek of the tenon to the wall of the mortise, and the lesser of the tensile strength of the wood in the dowels and the shear strength of the glue line around the dowels. The dowel joint is strongest when the outside dowels are as far apart as possible without getting so close to the end that the tenon is split by hydraulic pressure from the glue.

A mortise and tenon can be strengthened by increasing the size of the tenon, thereby increasing the glue area. The thinness of the glue line is also quite important—the thinner the better. The smoother the walls of the mortise and the sides of the tenons, the better the adhesion of glue to wood. I use a chain mortiser to make the mortise, and for the tenons, either a tenoning jig on the table saw or a single-end tenoner, which cuts with a cylindrical head like a jointer. But the tools used are not as important as getting a close fit.

While the decline and fall of Western civilization is widely anticipated, these things do take time, and until the event actually occurs there are few circumstances in which a door will be exposed to moisture other than that which is in the air. Therefore I generally use aliphatic resin (yellow) glue on doors that will be protected by a porch. This glue has worked out well in practice. In order to be classified "waterproof," a glue joint must withstand boiling water for some hours without losing strength. If you plan to boil your doors, phenol resorcinol glue is what you want. No matter what glue is used, be sure to seal both ends of the door with polyurethane varnish, even if the door is to be delivered unfinished. This is often neglected by the painter.

Wooden panels for a door can be flat or raised. Raised panels are somewhat easier to fit, because with flat panels the fit

Federal law requires manufacturers to use tempered glass, but permits leaded glass panels as long as no single piece of glass is larger *than 30 sq. in. and no opening is large enough to pass a baseball. Beveled octagonal glass, left, is a framed panel within a panel.*

must be precise—very nearly tight enough to split the stile or rail but not so tight as to actually do it. Something can be gained by slightly tapering the edge of a panel by a hand plane or belt sander but this requires a very light touch. Any irregularity or dip left by the plane will show up distinctly where the panel enters the frame. When using panels of glued-up stock, it is a good idea to design the door so that no panel is wider than about 12 in. This is particularly true where there is a cutout in the center for glass. If the panel and glass fit tightly, the wood of the panel may split at its narrow point when contracting, rather than moving in its grooves. Of course, don't fragment a design just to obtain narrow panels.

If a wide panel is necessary, flat-cut veneer over plywood will give great stability. Or large panels themselves can be made up as another frame within the frame of the rails and stiles, if the changing grain directions do not do violence to the design. A number of coats of polyurethane varnish on the door will inhibit the transfer of moisture from wood to air and reduce the shrinkage-expansion oscillations.

Often an integral part of the doors I design is a piece of stained glass that is curved or in some other way not rectangular. Installing the glass in the irregular opening can be a problem. The easiest solution is to let the glass into a groove when gluing up the door, in the same manner as for a wooden panel. This is quick and convenient, but impossible to repair. It is best to avoid this method unless the door is going to lead a quiet life in the interior of a mausoleum. Gentle curves can be glazed with moldings of steamed wood. First, a rabbet is cut with the router, then the glass is bedded in glazing compound, and finally the molding is steamed and put into place. I sometimes make a virtue of the necessity for fasteners to hold the molding and work brass screws into the de-

sign. Silicone caulk is excellent and long lasting, but it is also a glue and the window will have to be cut loose with a razor blade if it has to be removed. If curves are too acute for steam-bent wood, an extremely flexible brown plastic panel retainer can be used. It is available from the Woodworkers Supply Store, Rogers, Minn. 55374.

If neither steam bending nor plastics is appealing, you can use the band saw or sabre saw to cut a molding out of solid stock to fit the line exactly where glass and wood meet. This works well, but is time-consuming. Leave the stock 1/2 in. or more thick, make the cutout, then fashion some detail on the edge complementary to the sticking (the shape cut into the inside edge of the frame) on the door.

The sticking on all commercial doors is done so that the detail runs the full length of the stile. Its mirror image, called the cope cut, is then made on the shoulder of the tenon. The corner resulting when the door is assembled is a crisp line, much like that made by mitered molding.

The most economical way for a small shop to make these cuts is with matched coping and sticking cutters for the shaper. Knives can be purchased with standard copes and stickings already ground and many companies will grind a set to your specifications. I use a single-end tenoner with cope heads, which is somewhat more cumbersome to set up than the shaper but has the advantage of easily cutting a tenon as long as 3 in. and making the cope at the same time. Also, matching beading and coping bits are available for the router, and one could fashion a set of wooden hand planes to do the job. Skill and patience with hand tools can make a joint as well as a ton of machinery can, and also will lead one in the direction of simpler, less cluttered designs.

Relying on sticking to provide the detail on the inside edge of the frame works well if the panel design is rectangular and raised panels are used. However, when the design includes curved or flat panels, it is often better to eliminate crisp corners by cutting the sticking with a router after the frame is clamped up without the panels. The effect is to soften the corner, draw the eye away from the frame and emphasize the shape of the panels. Although subtle, the difference is important to the overall feeling of the door. Attention is diverted from the outline to the interior, for the most part unconsciously. Generally, soft corners are best suited to less formal designs, although this is not a hard-and-fast rule. Making use of this detail can be a powerful tool for the designer in trying to achieve a desired effect.

Moldings around the panels give a similar effect to conventionally cut sticking, but far more depth and detail are possible. The door can be made up with everything square and the moldings then glued into place. There is a problem here of wood movement, best solved by fastening the molding to the frame, leaving the panels free to expand and contract. Silicone-type glues will stretch a great deal while still holding their bond. Better yet, put the molding around the panel like a picture frame, with a channel or a tongue on its outer edge to fit to the door frame. No glue is needed to hold the panel or the molding in place.

Lately I have been experimenting with a molding that is H-shaped in cross section, with excellent results. The open ends of the H are cut to fit the stiles and rails on one side, and to fit the thickness of the panel on the other.

It is difficult for me to say anything really useful about design because only its superficial aspects can be discussed meaningfully. Much nonsense is spoken and written in an attempt to intellectualize style and lump it together with technique. More often than not, good design is a matter of trial and error combined with the designer's ability to recognize those combinations of color and form that succeed and, just as important, those that do not.

A number of design techniques, although they will not generate successful designs all by themselves, are nonetheless helpful from time to time. One of these techniques is to use a geometric form where possible rather than a free form.

Beveled glass takes on a multifaceted gemlike appearance when used in openings that are regular or irregular polygons. These same polygons around a free-flowing piece of stained glass give a visual reference that controls the curves on its interior. I suspect this explains why Art Nouveau was less successful in architecture than it was on a smaller scale. Its paintings were bounded by rectangular frames and its small objects and furniture by rectangular walls. Its architecture had no regular boundary and consequently appeared grotesque. Descriptions of space will go where they will but the human mind is Euclidean. And why should these geometric devices not succeed? Much of the diversity and beauty found in nature has as its foundation the geometric, crystalline structure of inorganic materials. A designer can do a lot worse than to mimic nature. At least it helps avoid appearing contrived.

Another interesting tool comes from the arithmetic series 0, 1, 1, 2, 3, 5, 8, 13, 21, 34, and so on. These are called Fibonacci numbers and each number in the series is the sum of the two preceding numbers. After the series has progressed for a while, the ratio between any two adjacent numbers stabilizes at 1.618. All this would be only of academic interest if someone had not noticed that the Parthenon fits neatly into a

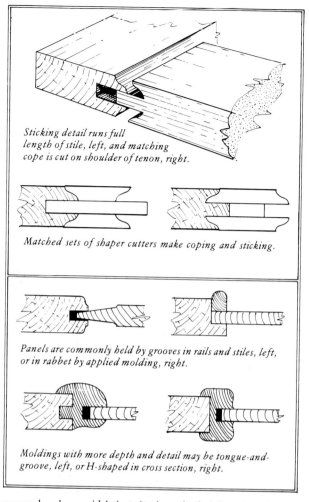

Sticking detail runs full length of stile, left, and matching cope is cut on shoulder of tenon, right.

Matched sets of shaper cutters make coping and sticking.

Panels are commonly held by grooves in rails and stiles, left, or in rabbet by applied molding, right.

Moldings with more depth and detail may be tongue-and-groove, left, or H-shaped in cross section, right.

rectangle whose width is 1.62 times its height; that the exquisite, logarithmically spiralled shell of the chambered nautilus can be generated with this ratio; that the proportions of some of Leonardo da Vinci's paintings, as well as those of Mondrian, seem to be determined by this ratio. A rectangle of this proportion, known as the golden rectangle, is frequently used in art and architecture. It has obvious applications to both doors and casework. Of course we do not want every rectangle to have these proportions, but it can be helpful to know the relationship.

These examples do not begin to scratch the surface. They are from one category of one mode of our awareness. That is, they are visual and oriented toward form. Within the visual mode there are also techniques for generating color and texture. And most often neglected are the other senses: smell, touch and hearing. The interplay and blending of techniques with a material as diverse in its nature as wood allows limitless possibilities for design.

And yet, when a door or piece of furniture succeeds, it is due to the designer's sensitivity rather than to manipulation and awareness of techniques. In much of the work where the golden rectangle has been found, the designer was unaware of the mathematics involved; the proportion just *looked* right. No doubt it is very easy to do a perfectly hideous piece based on the golden rectangle, or on any geometric figure for that matter. Techniques are just toys with which to play—they do not guarantee good design. Good design is simply done, not generated by formula.

☐

Paneled Doors and Walls
Colonial workmen relied on the right planes

by Norman L. Vandal

The earliest architectural application of bevel-edged paneling is found in wide-board wainscot, which covers or actually constructs the interior walls of our oldest houses. When applied to the interior surface of exterior walls, this wainscot was nailed horizontally to the vertical studs or planking which structurally supplemented the posts. Since the braced frame integrated with the chimney mass could support a second story, interior walls were never load-bearing, and in fact were structurally unnecessary. To divide chambers, and to provide a finished surface in both rooms, early housewrights used wide boards joined at the edges and fastened to floor and ceiling beams. These walls were from 1 in. to 1⅛ in. thick, the same dimension as the boards.

The conventional tongue-and-groove joint with the addition of a decorative bead was sometimes used to join such wainscot, but a beveled edge set into a groove on the adjoining panel appears to have been most common. For constant thickness and to make a wall with both surfaces in the same plane, the boards were beveled on both sides, resulting in a decorative feature visible on both sides of the wall. Feather-edged paneling is another name for such a wall treatment.

Panels joined by a simple tongue and groove had the undesirable habit of showing shrinkage as the two butted edges pulled apart. Feather-edged panels, having no distinct dividing line when viewed face-on, belied movement with the seasons, an important reason to use this type of joint. The beveled-edge joint was also used to construct interior and exterior plank doors. Wide boards were joined in the aforementioned manner and held together by clinch-nailed battens. Exterior doors were made of double-layered boards, the inside layer being horizontal and clinch-nailed through the vertical exterior layer. Original doors of this type are quite rare, time and weather having taken their toll.

In work dating from late in the 17th century we see many

(Text continued on page 62)

The paneled wall comes with its own vocabulary—one not found in modern dictionaries. Here are working definitions:

Fielding (or raising)—Sawing or planing a shoulder and a beveled edge all around a rectangular panel. The untouched center field thereby becomes raised.

Mullion—Vertical member within a framework. Also called a muntin, especially in windows.

Rail—Horizontal sticks of wood forming the top and bottom of a frame, to hold a panel. Rails are tenoned at both ends.

Stile—Vertical sticks of wood forming the outside edges of a frame that holds a panel. Stiles are mortised at both ends.

Frame—The rails and stiles, grooved on their inner edges, and assembled around a panel.

Stick—A long, narrow piece of wood suitable for making into rails and stiles. The process of planing a molding into the edge of a stick is called sticking, and so is the result. Such a molding is said to be stuck (as opposed to a separate molding nailed or glued in place).

Spring—The angle at which a plane is held to the stock.

Panel profiles

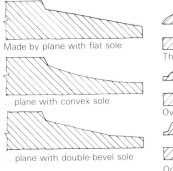

Made by plane with flat sole

...plane with convex sole

...plane with double-bevel sole

Rail and Stile profiles

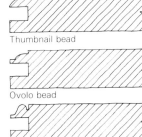

Thumbnail bead

Ovolo bead

Ogee

18th-century fireplace wall in a Massachusetts house; restored by author.

Kitchen was built and installed by author, during restoration of an 18th-century house.

Planes in use. Top. the A. Smith plane from previous photo; bottom, English panel plane. 3 in. wide by 8 in. long, has adjustable fences on sole and side for width and depth control.

Three panel-raising planes, circa 1800. Left. raising jack made by J. Butler of Philadelphia, has adjustable fence and depth stops, and nicker cutter ahead of skewed iron. This prevents tear-out when working across the grain. Plane at center was made by A. Smith of Rehoboth, Mass.; right was made by S. Kimball. All three planes are 14 in. long, and range from 2⅛ in. wide to 3¾ in. wide.

Molding plane top, and plow of unknown make (c. 1790), right. Plow has wooden screw locks to position slide arms, and sliding wooden depth stop. Planes of this type were usually sold with eight irons, for grooves ranging from ⅛ in. wide to 9/16 in. wide. Later planes usually had a closed handle for easier pushing, and screw arms with lock nuts.

Rare 18th-century combination planes, which cut grooves in rails and stiles and molded the edge in one operation. Plane at left, made by J. Woods, has two irons; plane at right, maker unknown, has a single iron.

A rare, three-part molding plane made by Isaac Field of Providence, R.I., between 1828 and 1857 should have a third iron, missing when the plane was found, fitted to the central section. When first discovered at auction, all three parts were held together with a wooden screw. The plane parts did not fit together well, and a little experimentation revealed that they were never meant to—the main body (right in the end view) was fastened to one or the other parts. Attached to the central section, the plane simultaneously plows a groove for paneling and sticks the molding on one face of rails and stiles. With the central piece removed and the two outside bodies attached together, the plane at once plows the groove and sticks the molding on both sides of rails and stiles, for doors with panels fielded on both sides. The groove is ⅛ in. wide and ⅜ in. deep, the moldings are both ⅜ in. thick. Plane from the collection of W. B. Steere. No. Kingstown, R.I.

EDITOR'S NOTE: The panel planes shown here are from the collection of Kenneth D. Roberts, who has written and published *Wooden Planes in 19th Century America,* vols. 1 and 2. He has also written extensively on the history of clockmaking, and publishes facsimile editions of old tool catalogs. For more information, write Roberts at Box 151, Fitzwilliam, N.H. 03447.

Joinery **61**

more raised-panel doors. A framework of mortised-and-tenoned rails and stiles formed the structure that housed the panels, which were beveled like the feather-edged boards. A groove was plowed into the framework's interior edges, into which the panels were fitted. On feather-edged boards a bevel was planed only upon edges running with the grain, whereas for raised panels, a bevel was also cut across the grain on the top and bottom edges. Many of the earliest doors had two wide panels, one above the other divided by a center rail. Later, four panels, two above two, became most common. The grain of the door panels always ran vertically, and the outside stiles, or vertical frame members, extended the full length of the door. Panels were never fastened, and were allowed to expand and contract freely yet invisibly.

On one side of a door, the inside edges of the rails and stiles were often given a decorative molding, mitered at the intersecting corners and forming a frame surrounding the raised panels. This treatment became the rule for most doors, despite the complications it caused in the making of the joints. Some doors were molded on both sides, but these are as scarce as they would have been difficult to make.

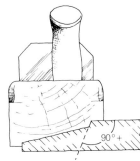

Sprung raising plane is held at 90° to bevel surface.

Regular raising plane is held at 90° to panel surface.

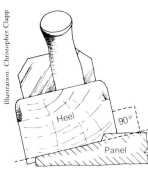

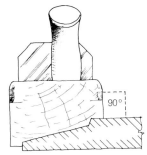

Shoulder angle is greater than 90°, thus the same plane can field panels of various thicknesses.

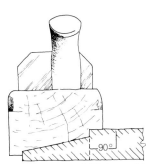

Shoulder angle of 90° is possible on planes used for a single panel thickness

...but on thicker stock such a plane undercuts the shoulder.

Near the end of the 17th century, wainscot on the surfaces of fireplace walls was replaced with sections of raised paneling joined in the same manner as the doors. An exception was the occasional use of an ungrooved framework of rails and stiles with rabbeted edges, on which bolection-type moldings were applied to house the raised panels. In all but rare examples, the decorative molding on the rails and stiles apparently was considered a necessary element of style. The back surfaces of raised-panel walls were never finished off unless they divided chambers and were visible in both rooms.

Toward the middle 18th century, raised-panel walls were nearly always of great beauty and style. Proportions were refined and the joinery was exacting. Walls adjoining the fireplace walls were sometimes sheathed entirely with wainscot to a height of about 3 ft.—the level of the window sills—and finished off to the ceilings with plaster. Stairways and entrance halls were embellished with paneling conforming to the geometrics created by the angles of the stair stringers. Raised-panel walls were integrated with ornate cupboards and fluted pilasters supporting elaborately molded cornices.

In the latter part of the Georgian period, the mantle with overmantle became the point of interest and portions of walls which were formerly paneled were now simply plastered. With the Federal period, raised paneling had all but disappeared. Even the overmantle was deleted as the fireplace surround became the focal point. Only the raised-panel door survived, its form and function being unsurpassed.

Panel-raising planes — Panels that were raised, or fielded, with raising planes exhibit two tell-tale characteristics. The shoulder of the beveled rabbet, where the feathered-edge meets the raised surface of the panel, is never perpendicular to the surface of the panel. Careful study of the soles and irons of planes will substantiate this fact.

Raising planes cut the shoulder and feathered the panel edge simultaneously. Because a single plane could be used to field panels of different thicknesses, the relationship between the angle of the bevel and the surface would have to vary. If the plane shoulder was designed to be at a right angle, on panels of heavy thickness the shoulder would be at an angle less than 90°, with obvious impractical consequences. Designing the plane to cut a shoulder bevel at greater than 90° avoids this problem. Of course the width of the groove also becomes a factor in the angle of the bevel.

Second, careful study reveals a convex sole on many planes, and a consequent concavity on the beveled edges of many panels. Not all planes have this feature, however, and it would be foolish to assume that panels with a flat bevel were not fielded with raising planes. On many planes, this convexity is actually a sole with two surfaces meeting at a shallow angle, its vertex where the panel would be fully seated within the groove of the rails and stiles. Planes of this sort, and those with a convex sole, have an advantage over the flat-soled type because the cabinetmaker can fit his panels to the groove with greater accuracy—the panel edge resembles a tongue more than a wedge.

Raising planes will seat themselves when the depth stop comes in contact with the panel surface, the cut being complete. However, the angle at which the plane is held to the stock will determine the thickness of the tapered edge. Scribing a line with a marking gauge and holding to it with the cut is one way to obtain consistent accuracy. A great many raising

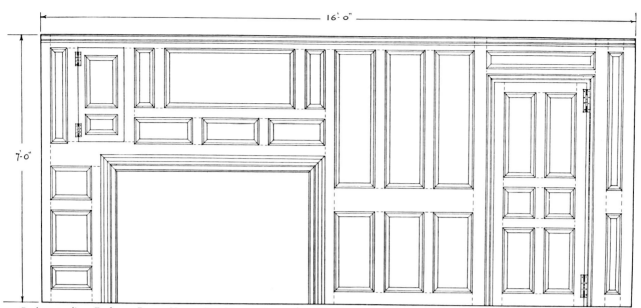

Parlor paneling, Forbes or Barnes House, East Haven, Conn., from The Domestic Architecture of Connecticut *by J. F. Kelly, Yale University Press, 1924 (Dover, 1963). The dotted lines, added by Vandal, indicate joints between rails, stiles and mullions.*

planes were made sprung, that is, they had to be held perpendicular to the bevel surface and not to the surface of the panel. This allowed the planemaker to make a tighter throat on the plane, greatly improving the smoothness of the cut.

Panel-raising planes, regardless of basic type, always had their irons set on a skew. This facilitated cutting across the grain on the ends of the panels. Planes that have a spur cutter or nicker preceding the iron are at an advantage over those without, as they pre-cut the shoulder and eliminate tearing across the grain. Planes with an adjustable fence allow the user a greater variety in panel-edge widths, but do not function any differently than the fixed-fence variety.

I can vividly remember questioning a friend's purchase of a raising plane, thinking my own method of using fillester and block planes quite effective in fielding panels. Combining efforts on a project with this same friend gave me the opportunity to use his plane. Its effectiveness was unquestionable, a special tool performing a very specific function.

The thickness of the edge of the panel could only be determined from the actual width and depth of the groove plowed into the rails and stiles. For this reason frame stock was probably gotten out first, then the panels made to fit the dimensions of the preassembled rails and stiles. Stock for the frames was grooved with either the conventional adjustable plow plane or with a plane similar to the groove plane in a set of match planes. After the frame stock was grooved or plowed, if a molding was to be used, it was now stuck with the proper plane. The grooving and molding combination plane eliminated this second step, as it could cut the molding along with the groove. The scarcity of this plane indicates that a two-step procedure using the plow and molding planes was most common. The 18th-century molding plane shown on page 61 (second box from top) has an ovolo bead configuration found on a number of Federal-period doors.

The molding was stuck upon the entire edge for the full length of the stock. The excess where the rails and stiles meet was later removed in the mitering and mortising process. Tenons were cut only upon the rails, and were of a length allowing them to pass through the mortises in the stiles.

Keeping the thickness of the tenon the same as the width of the groove facilitated making the mortise, but was not always the rule. In building a door, for example, the rails and stiles could be cut to dimension, the stiles being the total height and the rails the total width. The outside stiles, extreme vertical members of a panel framework, extended unbroken for the unit's full height. The two extreme horizontal members, the rails, extended unbroken between the outside stiles into which they were tenoned. Within this framework various arrangements of panels and framework could be devised.

Making panels — In reproducing raised-panel work, there are five basic principles necessary for historical accuracy. First, the cabinetmaker must extend his tenons entirely through the outside stile stock. Second, he must observe the rail and stile relationships mentioned in the previous paragraph. The discontinuity is immediately apparent when a raised-panel door (which was designed to hang vertically) is recycled as an overmantle. Let the doors remain doors. Third, he should field his panels in a manner resembling those fielded with a raising plane. Panels may be gotten out with various types of power machinery, but the time could be better spent making a raising plane which could be used repeatedly. Fourth, he must fully house the tenons within the stiles, and not cut corners by making deep slots in the ends of the stiles to accept the tenons. Doors made with slip joints can be pulled apart from top to bottom and are inferior in their ability to control seasonal movement. Fifth, he must not misinterpret the relationship between the tenon, the groove and the miter at the intersection of the molded edges.

Since the molding is stuck on the inside edge for the entire length of the frame members, that portion of molding where the stiles join the rails must be removed, as shown in the drawing on the next page. The molding on the stile is first mitered with a sharp chisel guided by a pre-mitered block clamped to the stock. The waste beyond the miter cut is then chiseled off or sawn off with a fine backsaw, so that the shoulder of the rail can butt against a flat edge on the stile. The molding of the rail must be mitered to fit the molding of

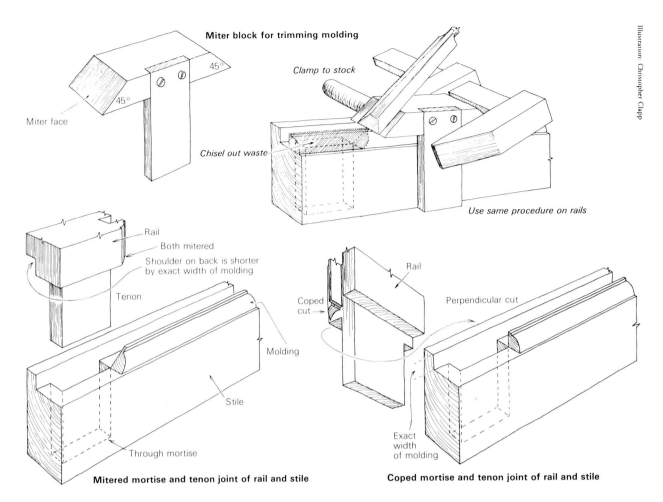

Miter block for trimming molding

45°

45°

Miter face

Clamp to stock

Chisel out waste

Use same procedure on rails

Rail

Both mitered

Shoulder on back is shorter by exact width of molding

Tenon

Molding

Stile

Through mortise

Mitered mortise and tenon joint of rail and stile

Rail

Coped cut

Perpendicular cut

Exact width of molding

Coped mortise and tenon joint of rail and stile

Illustration: Christopher Clapp

the stile. Where the molding is not contoured, as on the backs of the rail and stile above, it needn't be mitered, and a simple butt joint suffices. Simply cut the rail shoulder so that the back face of the tenon is longer from end to shoulder than the front face, by the exact width of the molding.

Instead of the miter, a coped joint was most frequently used to join the moldings at the intersection of rail and stile. Here one molding is cut away so that its profile exactly and neatly fits over the adjoining molding. As with the miter, the joints on the reverse side of the door or wall unit are simply butted. When viewed face-on, the coped joint looks the same as the mitered joint, and it sometimes takes careful study to differentiate. The coped joint has one superior feature: It overcomes separation of the miter due to cross-grain shrinkage. The coped portion moves over the adjoining molding like a slip joint. If doors or walls are painted during humid weather, the coped joint becomes apparent in dry weather, when shrinkage draws the coped portion away from the line of paint and reveals an unpainted section of molding.

Only frame members having a tenon, generally the rails and center or mullion stiles, are coped. The moldings on the mortised members, generally the stiles and where the mullion stiles fit into the rails, are not coped. They are cut off square and perpendicular to the length of the stock, flush to the ends of the mortises.

The coping is done by first mitering the molding, as on the fully mitered joint. The actual cope cut is made with small carving gouges at a right angle to the face stock along the curved line created where the contour of the molding meets

the plane of the miter cut. This line represents the shape of a molding of the same configuration coming into the coped molding at 90°. It is important to note that in using either the coped or mitered joint, the moldings stuck upon the edges of all frame members must be consistent. The coped portion of molding should fit snugly over the adjoining molding when the tenon is fully seated.

The pin holes, having already been bored through the stiles, are now used to locate the drilling points on the rail tenons. These are bored slightly closer to the shoulder of the tenon, so that the joint will be drawn together when the framework is pegged during final assembly. Pins are riven or split out and then whittled to approximate roundness. A slight taper is helpful in drawing the joint, and can also be easily whittled. The irregularities in the pin are enough to hold it within a perfectly round hole, bored slightly smaller than the rough diameter of the pin. Splitting and shaping pins require little effort, and scrap stock can be used.

In this day of power machinery and sophisticated glues and laminates, the art of getting out a raised-panel door or section of wall in the old manner has been forgotten by those outside of a circle of traditional craftspeople, knowledgeable tool collectors and old-house enthusiasts. It is a tribute to these people that the panel-making trade persists, to a limited but increasing degree. □

Norman Vandal, of Roxbury, Vt., builds and restores traditional houses during the summertime, and makes period furniture during the winter.

The Right Way to Hang a Door

by Tage Frid

When I make a door I first make the doorcase (frame). I make the inside of the doorcase 3/16 in. larger in height and width than the door itself. If the door is to be painted, I allow a little more for the paint. I bevel the edges of the door a little toward the closing side, so that if dirt or paint should fill up the corners of the frame, the door will still close tightly.

If I am using two hinges, I place them approximately one-sixth of the height of the door in to the center of the hinge. With three hinges, I center one and move the other two out closer to the top and bottom. The hinges should be mortised half into the door and half into the frame. Here is where the mortising plane comes in handy, as it is designed so that it will fit easily into the lip of the frame.

When fitting the hinges to the jamb, inlay the top hinge so that the door will fit tightly against the jamb when it is closed. But inlay the bottom hinge a little less, so there will be a gap of 1/8 in. or so in the back. Setting the hinges this way will leave the whole door cocked at a slight angle, which is much exaggerated in the drawing. The space will not be the same the whole way around. I do this because as a hinge starts wearing the door

will begin to droop down. Hinges set as described will allow for this droop and the door will fit much better throughout the life of the hinge. It is especially necessary to do this with modern stamped and rolled hinges. You can see in old doors that haven't been hung this way the extent to which drooping occurs.

When I install the door frame I have all the hardware—hinges, locks, latches—already installed in the frame. I use wood shingles as shims to level the frame and fasten the hinge side first. Then I hang the door into the frame, close it and shim it until I get the spacing I want all the way around the door. Then I fasten the rest of the frame to the studs on the wall.

If I really want to do it right, I use a door-frame dovetail on all four corners, with the pins in the horizontal pieces. The joint is designed so that when you fit the door you can make the frame narrower or wider without a gap showing. Also, if the door should shrink or expand, I can take the outside molding off and wedge in or shim out the door frame to fit the door without getting a gap and without having to plane the door and refit the hardware. And this joint is much stronger than the usual method of nailing the corners together. □

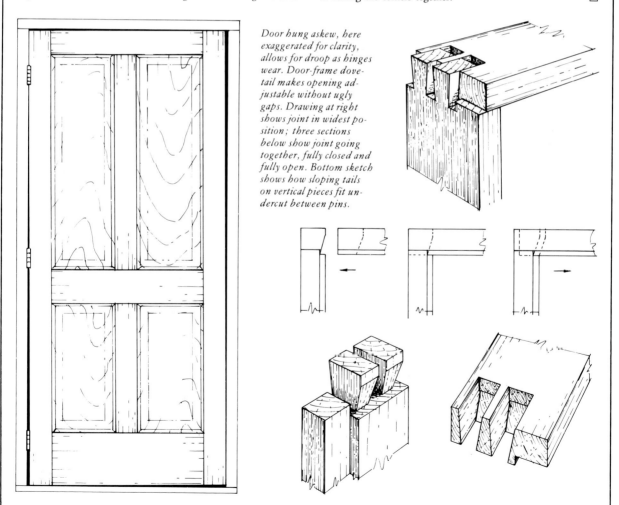

Door hung askew, here exaggerated for clarity, allows for droop as hinges wear. Door-frame dovetail makes opening adjustable without ugly gaps. Drawing at right shows joint in widest position; three sections below show joint going together, fully closed and fully open. Bottom sketch shows how sloping tails on vertical pieces fit undercut between pins.

Tambours in the Weed kitchen allow cabinets (all made of black willow) to remain open without swinging doors on which to bump your head or to obstruct access. Weed says this design requires no more work to make than a set of frame-and-panel doors. Right, tambour curtain disappears behind faceboard and false back, painted white. A stiffener strip neatens the top edge of the false back. Cabinet is deep enough for stacks of 10-in. plates.

Tambour Kitchen Cabinets
The conveniently disappearing door

by Richard Starr

"These doors are incomparable as far as convenience is concerned," says Hazel Weed about the kitchen cabinets built by her husband, Walker. "When I'm in here working, I just open them up and have everything where I can get at it." Weed, a woodworker who directs the student craft shops at Dartmouth College, feels that traditional doors are always in the way, obstructing other cabinets and banging you on the head. When planning the kitchen for their old farmhouse in the hills above Hanover, N.H., the Weeds chose disappearing doors—vertically opening tambours.

Most cabinets are boxes with doors hung on them, but a tambour must be designed into the cabinet structure. Weed routed the curtain track into the inner sides of the carcase (front, top and back) before assembly; the three straightaways were joined by tight curves. To ensure that the left and right tracks were identical, Weed used a template to guide the router. A false back, hung in grooves in the sides and trimmed with a stiffener strip at the top, hides the back of the open door from view. To cover the curve of the door as it arches into the cabinet, Weed installed a faceboard whose inner surface was contoured to match the track. Thus the vis-

ible section of the door is flat, creating the illusion that it is disappearing into the air over the cabinet as it is opened.

For the curtain to run properly the cabinet must be perfectly square; Weed emphasizes the need for accurate joinery. He fixed the cabinets to the wall only at the top so they would not be distorted by any settling of the house. The carcases are joined with half-blind dovetails at the corners. The dividers in the double and triple cabinets are set in dadoes; their front ends are housed dovetails that extend about 1½ in. back into the cabinet.

To make the curtains, Weed cut a board slightly longer than the width of the cabinet and planed it to ½ in. thick, the width of the strips. He rounded the edge of the board with a router, then ripped off this half-round strip, ½ in. thick, on the table saw. The newly sawn edge of the board was shaped again and the process repeated until he had enough strips.

The finished tambour curtain must be long enough to occupy both curves of the track when closed; its width will be slightly narrower than the distance from one groove bottom to the other. After washing and ironing the 14-oz. canvas backing, Weed cut it to a width slightly less than the inside of

Photos: Richard Starr; drawing: Joe Esposito

Tambour kitchen cabinet

Overall dimensions:
30 in. high by 36 in. wide
by 11¼ in. deep

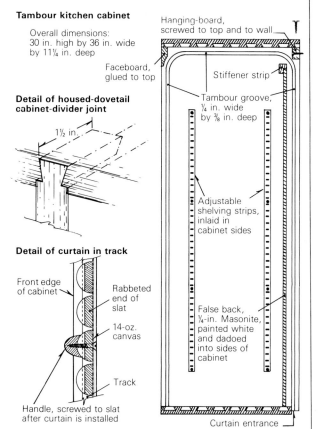

Hanging-board,
screwed to top and to wall

Faceboard,
glued to top

Stiffener strip

Tambour groove,
¼ in. wide
by ⅜ in. deep

Adjustable
shelving strips,
inlaid in
cabinet sides

False back,
¼-in. Masonite,
painted white
and dadoed
into sides of
cabinet

Curtain entrance

Detail of housed-dovetail cabinet-divider joint

1½ in.

Detail of curtain in track

Front edge
of cabinet

Rabbeted
end of
slat

14-oz.
canvas

Track

Handle, screwed to slat
after curtain is installed

---

the cabinet, not counting the track. He left the canvas longer than the finished curtain, and trimmed it after gluing. Each strip was fixed to the cloth with a thin line of Titebond glue down its back, its length centered on the width of the fabric. The wood strips were butted closely together and weighted under plywood to dry. Weed suggests running the curtain over the edge of a table to break the hardened oozed glue, then folding it back on itself to scrape excess dry glue from the edges of the strips. The curtain was trimmed to width, then its edges were rabbeted on the table saw using a batten to hold it down on the saw table. Weed compares this step to fitting a drawer: Both surfaces of the rabbet must fit the groove closely but there must also be sufficient clearance.

After the cabinet is hung, the curtain slides into the track from the back end. Weed suggests waxing the rabbet before installation to ensure smooth running. Some friction is desirable since the tambour is perfectly counterbalanced only when half open; when more or less than half open, it wants to slide up or down. In actuality, the curtain stays put everywhere but a few inches from the ends of its travel. With the curtain in place, a fitted handle was screwed to the predrilled second slat. The handle stops the curtain against the facing at the top of the cabinet.

Weed estimates that a set of tambour kitchen cabinets would take no more time to make than a set with well-crafted frame-and-panel doors. The function of cabinet doors is to hide the clutter and to keep out the dust. Vertical tambours do it efficiently and beautifully. □

Richard Starr, author of the book Woodworking With Kids *(Taunton, 1982), lives in Thetford Center, Vt.*

Q & A

Plate joinery—Here's a method for cutting the slots for La-mello-type joining plates without having to buy the expensive hand power tool. We simply bought one of the carbide-tipped replacement blades and mounted it on our radial-arm saw. The blade's arbor hole is ⅞ in., so we machined a bushing out of a steel washer to match it to the ⅝-in. saw arbor. We tilt the motor to the vertical position and cut the slots using the jig shown below.

The jig includes a slotted fence that clamps in place of the saw's regular fence, and an auxiliary table that raises the work high enough so that the arbor nut will clear. The slot is positioned by referring to index marks on the fence; a stop block clamped in the saw track controls its depth. Make sure the blade arc is parallel to the table, or your finished parts will have a twist. To slot the end of a rail or a stile, we clamp another fence at right angles to the slotted fence and make the usual cuts. —*Bernard Theiss and Linda Fisher, Sharon, Pa.*

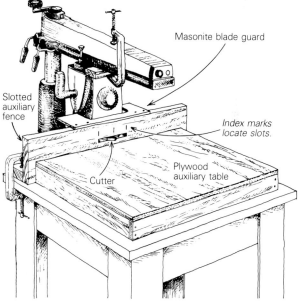

Masonite blade guard

Slotted
auxiliary
fence

Index marks
locate slots.

Cutter

Plywood
auxiliary table

End-to-end gluing up—*I'm planning to build an 11-ft. counter out of 6/4 white oak, but I don't have boards long enough. Is it acceptable to glue up random-length lumber end to end to make an 11-ft. counter?*
—*Patrick Warner, Escondido, Calif.*

TAGE FRID REPLIES: Yes, you can glue up your boards end to end to make them long enough. First square the ends of the boards and cut ¼-in. grooves in the end grain with a slotting cutter in a router, if you have one. So the splines will be strong and to keep seasonal movement from cracking the boards at the joint, run the grain in the splines in the same direction as the grain in the boards. Glue your boards lengthwise before you rip them to final width. If you don't have

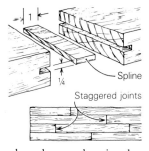

1

¼

Spline

Staggered joints

clamps long enough, just nail blocks to the floor at each end of the stock and apply pressure to the joint with wedges against the blocks. When you edge-glue your lengthened boards into the finished counter, stagger the joints to lessen the stress on each one. I'd suggest you use a hard film finish such as polyurethane; otherwise the tannic acids in oak may react with metal that comes in contact with the counter, staining it.

Solid Wood Doors
How to make them and keep them flat

by Tage Frid

There is no way to stop solid wood from moving, except to make it into plywood or to treat it chemically, which kills its color and beauty. When making anything out of solid wood, you must be sure to control which way the wood moves, and leave room for it to do so. I have seen a ½-in. thick parquet floor, when the roof started leaking and the floor got wet, push out all four brick walls about six inches.

The problem of wood movement is complicated when we want to make doors out of solid wood for houses or cabinets. We want the door to stay flat, and to always fill its opening. But not only does the wood expand and contract across the grain, it also tends to bow. This is because the humidity on each side of the door is usually different, especially when the door separates the outdoors from the inside of a house. Even a cabinet door is exposed on its outside to various temperature and humidity changes in the room, while very little changes

inside the cabinet. If the door is not constructed to stay straight it will bow out when humidity is high in the room, like during the summer, and cave in during the winter when the heat is on. Usually the door will also try to twist.

Although a cabinet door is seen on both sides, it is usually closed and the inside exposed only when it is opened. Since it does not affect the door's function, I would put the more beautiful side of the wood toward the outside. Some textbooks suggest that you alternate the cup of the annual rings; however, there is no need to do so. The wood can be kept straight with one of the methods I am about to describe. Whatever way you choose, if it is done right and made well, it will add to the design. Anything that is constructed and designed right is beautiful because it makes sense. The outside of an airplane was never designed to be beautiful, but because of the way it slips through the air, it is beautiful.

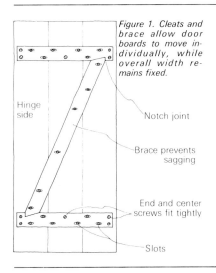

Figure 1. Cleats and brace allow door boards to move individually, while overall width remains fixed.

Hinge side

Notch joint

Brace prevents sagging

End and center screws fit tightly

Slots

When you are making a solid wood door for a house, especially an outside door, don't glue the boards together. Construct the door so that its two outside edges are tied down. The individual boards will be able to move, but the overall width of the door won't be able to change. There are several good ways to do this; the oldest and best-known method is shown in figure 1. This is the way most people would make a solid wood door and there is nothing wrong with this construction.

The diagonal brace and cleats are what keep the door together. It is important to fit the brace into the top and bottom cleats with a notch joint, as shown, so it cannot move. Also be sure that the brace is anchored to the bottom cleat on the hinge side, and to the top cleat on the lock side.

This arrangement prevents the door from sagging. If you change the hinges to the other side, turn the door upside down.

The brace and cleats are held to the door with screws. The screws at the ends of the cleats should fit snugly into their holes, so the total width of the door cannot change. In this example the center board is also tied down in its center, so the wood can move an equal amount on each side. All the other screw holes in the cleats and brace are horizontal slots. When the boards move, the screws can move back and forth with them. If you don't do this the boards will split. The length of the slots depends on the width of the board. Using a drill a little larger than the screws, drill three holes side by side. Remove the waste with a chisel or keyhole saw.

Figure 2A. Nailed batten can cover the gap in tongue-and-groove doors, which work best when the boards are narrow.

Batten

Figure 2B. Lap joint can accommodate movement of wider boards. Slots in cleats permit screws to slide.

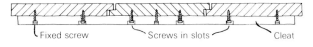

Fixed screw   Screws in slots   Cleat

Figure 2C. Two layers of wood nailed or screwed together with staggered joints make a tighter door.

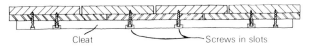

Cleat   Screws in slots

The door must also be wind and rainproof. The most common way to do this is tongue-and-groove. This is fine with narrow boards, but the door won't stay airtight if the boards are wide. The tongue usually isn't very big, its edges are often rounded and sometimes the tongue is tapered so the joint will slide together easily. The result is that when the boards shrink, wind and rain come right through. You could nail batten strips over the joint (figure 2A). With wide boards, a lap joint (figure 2B) can be quite successful, but I prefer to use two layers of wood (figure 2C) for a tighter door. The boards can be screwed together, or if the wood is not too hard, nailed together and the ends of the nails bent over. Just be sure the ends of the cleat are screwed down tight, as before.

When using strap hinges, the back of the door is exactly as in figure 1, but the cleats and brace are welded steel. The hinges are bolted to the front of the door, through the wood and the steel on the back side. Usually there is only one bolt through each board. If the boards are not too wide, being bolted between the two pieces of steel overcomes warping. If more than one bolt has to go through each board, some of the holes in the wood have to become slots.

Figure 3. Steel rod through boards can overcome wood movement.

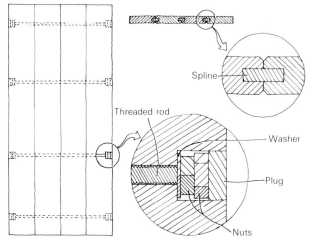

Spline

Threaded rod

Washer

Plug

Nuts

Figure 3 shows a door without cleats or strap hinges, but with metal rods right through the width of the boards. The best way to make a door like this is to use cold-rolled steel threaded at both ends, or threaded rods. To make the door as tight as possible, put the boards together using splines, but don't glue them together. The

bolt holes through each board have to line up. First, drill the big holes for the nuts and plugs in the outside edge of board no. 1. Then use a long drill of the same diameter as the rod and continue the holes all the way through. Be sure the holes are centered and straight. Drill only a little at a time then pull out the bit and clean it, or else the drill will start wandering. Most hardware stores stock 12-in. and 18-in. bits that electricians use, called bell-hanger's bits. The most common size is ⅜ in., and this is a good size for the rod too. For a larger door, or an outside door, use ½-in. or bigger rod. With all the holes drilled in the first board, put the spline in and clamp the first and second boards together. Use the long bit and the hole in the first board as a guide to drill partway into the second, then remove no. 1 and continue drilling all the way through. Proceed in the same way through nos. 3 and 4. Now to accommodate the nuts and plugs on the outside edge of no. 4, plug the rod hole with a dowel (don't glue it) and drill the big holes.

With all the holes drilled, assemble the door and insert the rods, then put the washer and one nut on each end. Tighten them up with a socket wrench so they are good and tight, but don't overdo it by compressing the wood. It will compress anyway, when the wood expands, but if you do it right the boards will stay about the same width after the wood shrinks again. Any small gap will remain airtight because of the spline. Put a second nut on to lock the assembly, then glue in the plug. Square one end of the door and cut it to length. With the rods in, the width of the door will always be the same even though each board can move a little bit. Be sure not to put the rods where the lock and hinges are going to be.

Figure 4. Breadboard ends allow wood to move both ways—remember to allow for expansion as well as contraction.

Figure 5. Cleat with sliding dovetail will hold a cabinet door flat.

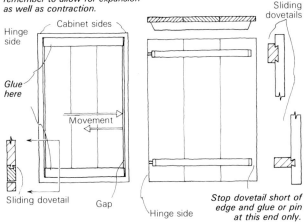

Cabinet sides

Hinge side

Glue here

Movement

Sliding dovetail

Gap

Hinge side

Sliding dovetails

Stop dovetail short of edge and glue or pin at this end only.

Attaching two pieces of wood to the top and bottom of a cabinet door (figure 4) is another way to keep a door straight. The end pieces should be fastened with sliding dovetails and should be secured toward the hinge side. This way, all the expansion will take place in the opposite direction, but you have to remember to allow for it. If you were to secure the sliding dovetails at the center, like a breadboard, the hinges would be pushed out when the door shrinks while it is enclosed between the two sides of the cabinet. I try to avoid this construction for solid wood doors whenever possible because it looks like half of a frame-and-panel without any of the advantages of a frame-and-panel.

For a cabinet door or the top for a chest where the boards are glued together, cleats could be slot-screwed on the inside. But for a piece of furniture, I would attach the cleat with a sliding dovetail instead. The wood for the cleats should be on edge against the door—for stiffness—but it also could be on its face to be less obtrusive. Stop the female dovetail housing near the handle side, so it doesn't show on the edge when the door is open, and glue the dovetail only toward this end. Thus the wood will be free to move. This is shown in figure 5.

Figure 6. Deep cleats can become the top and bottom of shelves mounted inside the cabinet door, like a refrigerator. Dovetail the vertical sides of the door shelves to the cleats, and add a front rail to keep things from falling out when the door is opened.

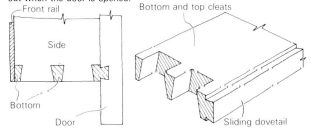

Front rail

Side

Bottom

Door

Bottom and top cleats

Sliding dovetail

Figure 7. Movement can be concealed by letting the door overhang the cabinet...

...or by leaving a gap and concealing it with an applied handle.

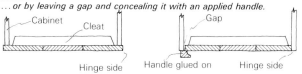

Cabinet

Cleat

Hinge side

Gap

Handle glued on

Hinge side

When you use a cleat with a sliding dovetail on a cabinet, the inside of the door can become a shelf, like on a refrigerator door. Make the cleats wide enough for what you want to store, and dovetail their ends to vertical side pieces, as shown. Be sure that you glue a piece of wood in front of the door shelves, or join on a railing, to prevent what is stored there from falling off when the door is opened (figure 6).

Because such a door will always move in width, you have to make allowance for movement when putting the hinges on (figure 7). One way is to mount the door outside the cabinet, with an overhang all around. If the door is set inside the cabinet opening, you must leave space for expansion along the edge opposite the hinges. One way to conceal the gap is to glue a piece of wood onto the front edge of the door, and shape it into a handle.

Today you have to be lucky to find a board that is wide enough to make a door by itself. But if you are so lucky, treat the board just as if it were several boards glued together. Always allow solid wood room to move. ☐

Joinery Along Curved Lines
A general method for template routing

by Jim Sweeney

*Bearing on router-bit shank
solves curve-tracing problem.*

We woodworkers are fascinated by joinery. Often we'll examine a piece of furniture to see how it's put together, even before we consider how it looks or functions. Exposed joinery is particularly satisfying. Through its visible joinery, a piece speaks of the fact that it is wood, and of all the possibilities and limitations of assembly which that implies. In this article I'll discuss a general method for extending exposed joinery from the ordinary linear joint to joints that follow almost any arbitrary curve. Using only a router and a few combinations of straight cutters and bearings, complex but precise curved joints become available to anyone willing to apply some ingenuity.

In making furniture with non-traditional shapes, it's particularly valuable to have a vocabulary of curved, exposed joints that are analogous to the lap, bridle and dovetail of rectilinear joinery. The principal lines in contemporary furniture often are not straight; the maker tries for a coherent whole using the rhythms and tensions of curved lines and planes. But in a non-traditional shape, as in an unrhymed poem, the sense of rightness is elusive. The eye is less forgiv-

ing, less able to fill in, when it is faced with the unfamiliar. The designer of unfamiliar shapes struggles against the pull of the amorphous, the chaotic, or the simply awkward. In this context, traditional joint shapes can be especially inappropriate, unless used in irony or in contrast—they are words from another language. Of course, these problems can be avoided by using concealed joinery (mortise and tenon, dowel, etc.) or non-joinery (stack lamination), but such choices just eliminate one more option in resolving the design.

It's not surprising that the tool used to cut curving joints is the router. The other joinery tools that have evolved—saw, chisel, plane, and their machine counterparts—cut flat planes. The router, unlike these older tools, does not need to refer to where it has just been to determine where it is going. At any instant it is free to move exactly where we guide it. The problem of curved joinery thus becomes how to guide the router to cut the precisely matching negative and positive curves that will unite to form some particular joint.

The main tactic for doing this is to shape templates of some durable material (plywood, Masonite, aluminum) and to transfer the shape to the work, by using a straight two-flute router bit with a ball-bearing mounted above its flutes. There are thus three parts to the problem of cutting curvilinear joints: first, how to make precisely matching negative and positive templates of a chosen shape; second, how to transfer these shapes into the workpiece; and third, how to make this transfer occur exactly where we want it in the finished work. Though this third aspect of the problem may at first seem trivial, it often calls for the biggest bag of tricks, and is the least susceptible to any kind of general solution.

Let's first consider the second problem, that of transferring the curve from a template to the work, because it best introduces the use of bearing-cutter assemblies. My technique descends directly from two common techniques, so I'll start by describing them and their disadvantages.

In one standard technique the router is guided, as it is fed through the cut, by its subbase being pressed against a template or fence that is clamped to the work. In a more flexible version of this method, a template guide-collar is attached to the subbase of the router and run against the template, while the router subbase rides atop the template. The cut may go completely through the work, or only partly into it. But in either case there is one problem: concentricity, or rather, the lack of it. The router bit is reasonably concentric with the motor, the motor is somewhat loosely housed in the base, the subbase is aligned with the base by some hopeful screws, and finally, the guide collar fits into a hole in the subbase. All of these attachments are governed by the initial imprecision of manufacturing as well as by subsequent wear. We should not be surprised when this chain of vagueness does not yield pre-

*Bridle joint with
curved shoulder.*

cise concentricity. The cutter changes position whenever the router is rotated relative to the template. So the worker must try to keep the same point of the subbase, or guide collar, against the template throughout the cut—a hopeless task when so much else compels his attention.

In a second standard method, the guiding surface is built into the cutter, so that concentricity is guaranteed and overall accuracy is limited only by the precision of the cutter assembly itself. This technique employs a flush trimmer bit—commonly, a $\frac{1}{2}$-in. straight cutter with a $\frac{1}{2}$-in. ball-bearing at its end—with the bearing rolling against a template that is attached to the work on the face opposite the router. This method is limited mainly by the position of the bearing: we can trace through work only as thick as the cutting flutes are long, and we must always cut completely through the work. The first limitation is not serious because you can find trimmer bits as long as 2 in., though they are fragile. The second limitation, always cutting through the work, is very restrictive—a mortise cut this way, for example, must always be a through mortise. You cannot use a trimmer bit to cut the negative part of a half-lap or a bridle joint.

Furthermore, trimmer bits themselves contain a subtle source of error. Because they are designed so that the bearing can roll against a vertical, Formica-covered surface while the cutter trims the adjacent horizontal surface, the diameter of the cutting circle is usually 0.005 in. to 0.010 in. smaller than the bearing. This minute difference can become significant when you make a template from a series of intermediate templates, because the errors accumulate. You can sometimes compensate with a layer of tape on the template.

There is another technique for transferring a curve from template to workpiece, but it requires a pin router. A template fixed to the underside of the workpiece can run against a pin the same diameter as, and directly below, a bit entering the workpiece from above. The entire edge of the workpiece can thus be shaped to your curved pattern (see page 78).

The system I consider optimum combines the best features of both standard techniques: a bearing on the bit for concentricity, but placed on the router side of the cutter, on the shank above, rather than below, the cutting flutes. This bearing will follow a template exactly, yet the cut can stop at any desired depth. The outside diameter of the bearing is ordinarily the same as the cutting diameter, but it can also be larger. This simple idea, it turns out, unlocks a host of joint designs that can be as simple or as complex as the craftsman wishes. Any pair of fitted shapes that you can make in $\frac{1}{4}$-in. hardboard can be transferred into solid wood, either completely through the workpiece or to any depth, although it is sometimes necessary to juggle template thickness and cutter length.

Router bits with bearings mounted on their shanks, unfortunately, are not stock items. And although router bits are made in a wide range of sizes, ball-bearings are not. The woodworker must devise bearing/bit combinations that will do the work at hand, and find a mill supply house or industrial hardware dealer who stocks or can get the parts. For

Author's open-sided chest-of-drawers relies upon curved bridle joint for visual continuity. Chest is under construction in photo below, and finished in detail on facing page. Scalloped template (shown standing at right) straddles the chest's vertical spine to rout the curved shoulders of the bridle joint. A mating template attaches to the horizontal arms that hold the drawers, and the ends of the arms are routed to fit the spine. Inset drawer pulls are shaped with the same template techniques; a similar handle is explained in detail on pages 74 and 75.

most of my work, and in the examples that follow, I use either of two bit/bearing combinations: a $\frac{3}{4}$-in. diameter cutter with $\frac{3}{8}$-in. diameter shank plus a bearing of $\frac{3}{8}$-in. ID and $\frac{3}{4}$-in. OD (for example, New Departures' #77R6AB), or a $1\frac{1}{8}$-in. diameter cutter with $\frac{1}{2}$-in. shaft plus a bearing of $\frac{1}{2}$-in. ID and $1\frac{1}{8}$-in. OD (for example, TRW's MRC R877.)

The only limitation here is that the smallest radius of curvature in the joint must equal or exceed the radius of the smallest bearing. With my $\frac{3}{4}$-in. combination, a $\frac{3}{4}$-in. diameter washer must be able to roll along the curve and touch every point of it. You can get around even this limitation by using unmatched bearing/bit combinations, $\frac{3}{4}$-in. to $\frac{3}{8}$-in. for example, with appropriate offset in the template.

Usually the bearings simply slide onto the bit shank until they are stopped by the metal of the cutting flutes. If there is any resistance, apply force to only the inner race of the bearing. In extreme cases, expanding the bearing with heat and

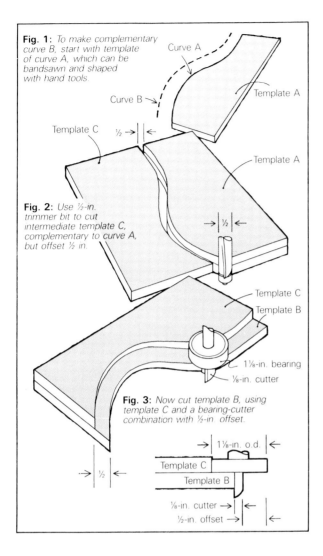

Fig. 1: *To make complementary curve B, start with template of curve A, which can be bandsawn and shaped with hand tools.*

Curve A

Curve B

Template A

Template C

Template A

½

Fig. 2: *Use ½-in. trimmer bit to cut intermediate template C, complementary to curve A, but offset ½ in.*

½

Template C

Template B

1⅛-in. bearing

⅛-in. cutter

Fig. 3: *Now cut template B, using template C and a bearing-cutter combination with ½-in offset.*

1⅛-in. o.d.

Template C

Template B

½

⅛-in. cutter

½-in. offset

contracting the shaft by freezing will help. On the router end, the collet itself stops the bearing from sliding up on the bit shank. A drop of Loctite will hold it fast.

Of my two bearing/cutter combinations, the smaller is best for inside cuts (that is, when the bit is surrounded by wood), where control and power are most important. The larger is for outside cuts, where good surface finish and freedom from tear-out are necessary. Bear in mind, though, that even a powerful router is an ineffective hogging tool, because the counterforces developed in a heavy cut make precise control difficult. Whenever possible, it's best to use the router as a trimming tool, with some other tool—drill press, gouge, dado blade, or bandsaw—removing the bulk of the waste.

Now that we have a way to trace a template curve down into the workpiece, let's return to the original problem of fitting two pieces of wood together along some curve—that is, how to create two matching templates, one negative, the other positive. Let's assume we already have an arbitrary curve A and we wish to create the matching curve B (figure 1).

Clamp or tack a piece of template material on top of template A, leaving plenty of this template material on the off side of the curve. We will cut an intermediate curve, C, which will be exactly like the desired curve B except offset ½ in.—too small or too large, depending on whether it is concave or convex. We can do this by tracing along curve A with

an accurate ½-in. trimmer bit, as though creating a copy of A. What we want, however, is the fall-off (figure 2).

This is the only stage in the process where we are at risk, because any deviation by the bearing from perfect contact with A will result in a defect in curve C, which will be inherited by curve B. You might need several tries to get it right, but the price of failure is low—a wasted piece of hardboard.

Now, to obtain curve B it's necessary to cut a curve parallel to C but offset ½ in. in the other direction. Tack C onto a piece of template material, and cut out B by rolling along C with an unmatched bearing/cutter combination, one with a bearing 1 in. larger in diameter than the cutter. One such combination would be a 1⅛-in. diameter bearing on a bit with a ½-in. shank and ⅛-in. cutting diameter (figure 3).

Although in theory this would work perfectly, in practice ⅛-in. diameter cutters are fragile and easily broken. I therefore had a machinist turn a bushing with a 1⅛-in. ID and 1¼-in. OD, and press it onto a ¼-in. ID by 1⅛-in. OD bearing. This maneuver created a 1¼-in. bearing that I could mount on a ¼-in. cutter. The difference in radii is the desired ½ in. Another way to achieve the transition from C to B is to make two setups with a 1⅛-in. bearing on a ⅝-in. cutter, first using C to cut an intermediate curve, D, and then using D to create B. This tactic achieves a ½-in. displacement by adding two ¼-in. displacements. A number of other combinations will also work as well.

Curve B should fit precisely into curve A—unless, of course, something went amiss, which it frequently does. One source of error is any difference between the diameter of the cutter and the diameter of its bearing, likely if the bit has ever been sharpened. A discrepancy can be corrected by applying tape along curve A, as close as possible in thickness to half of the difference between cutter diameter and bearing diameter. Masking tape, packing tape, duct tape, and ordinary Scotch tape all have usefully different thicknesses.

Before going on to consider the third part of the curve-fitting problem, that of using these templates to rout our matching curves exactly where we want them, I should mention a few ways to generate curve A. Bandsawing and hand-sanding work, though it's difficult to maintain a perpendicular edge. If the edge isn't perpendicular to the face of the template, you have not one but several curves, all vaguely close to A, depending upon which cross-section the bearing rides. A drill press with sanding drum, or better still a spindle sander, is more effective. A good way of generating curve A in the first place is to combine several sub-curves already on hand, for example circles of various diameters (cut with lathe or drill press), straight lines, or french curves, tracing out the various components one at a time with a bearing and bit, all the while maintaining fairness where they merge and overlap.

To the third problem, that of positioning the matching curves in the work, I can give no general solution. Each joint requires its own tricks. Perhaps it will be useful to discuss a joint, the trick for which occurred to me suddenly as I was riding on top of a crowded bus to a dive in Moorea, dodging coconut fronds, and certainly not thinking about woodworking. Suppose we have templates A and B, with which to cut boards I and II, so that they may be joined in a lap joint, as shown in figure 4.

A template can be aligned on board I by making the respective edges flush and by making some point on the curve coincide with some point on the edge. But board II has no

Drawings: Jennifer Kallgren

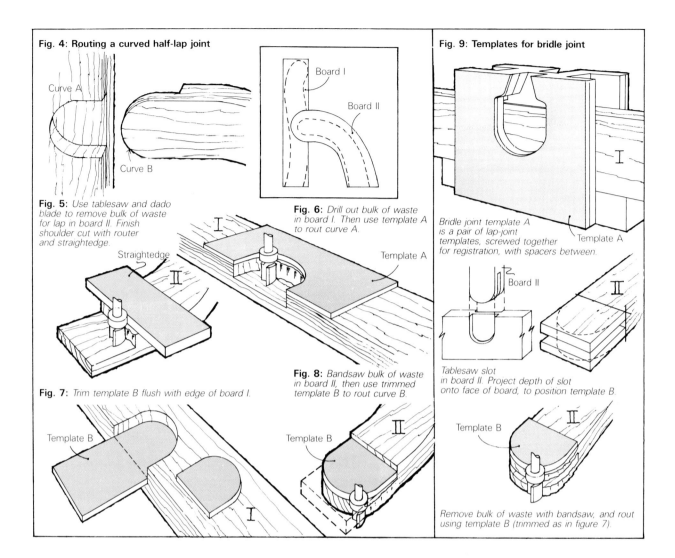

Fig. 4: Routing a curved half-lap joint

Curve A

Curve B

Board I

Board II

Fig. 5: *Use tablesaw and dado blade to remove bulk of waste for lap in board II. Finish shoulder cut with router and straightedge.*

Straightedge

II

Fig. 6: *Drill out bulk of waste in board I. Then use template A to rout curve A.*

I

Template A

Fig. 7: *Trim template B flush with edge of board I.*

Template B

I

Fig. 8: *Bandsaw bulk of waste in board II, then use trimmed template B to rout curve B.*

Template B

II

Fig. 9: Templates for bridle joint

I

Template A

Bridle joint template A is a pair of lap-joint templates, screwed together for registration, with spacers between.

Board II

II

Tablesaw slot in board II. Project depth of slot onto face of board, to position template B.

Template B

II

Remove bulk of waste with bandsaw, and rout using template B (trimmed as in figure 7).

straight edges, and it will be difficult to position the template to cut a half lap so that curve B has the same relationship to the lap's shoulder as curve A has to board I's edge. Do we cut the lap first, or curve B first? Proper cutting sequence is crucial in curved joinery, because each curve cut eliminates one more straight line from which to index.

In this case, it makes sense to cut the lap shoulder first. Start by scribing a line on board II for the lap shoulder. This doesn't have to be exact, as there is some allowance for waste, but the closer it is, the less waste there need be. Rough out as much wood as possible with tablesaw and dado blade, cutting to the finished depth but not to the lap shoulder line. Because of the curvature of board II, it would be awkward, though not impossible, to cut this shoulder on the saw. Instead we turn the work over and use a router with a straight-edge template to finish the shoulder (figure 5).

Now mark out curve A on board I, and drill out most of the waste. Clamp template A to board I, lining up the edges, and cut curve A into board I with the router (figure 6).

Now, and this is the crucial trick, slide template B into the A-shaped cutout you just made in board I, and tack it tightly. Then trim template B flush with the edge of board I (that is, the edge that ultimately will butt against the lap shoulder in board II). Use either an accurate trimmer bit, with the router base riding on the cut-out face of board I, or a bit-

with-bearing-above-cutter, with the router base riding on the face opposite the cutout (figure 7).

Tack this truncated version of template B tightly against the lap shoulder in board II. The lap shoulder of board II will fit against the edge of board I exactly as tightly or as loosely as template B fits against the shoulder. Cut template B's shape into board II using a cutter with same-size bearing above its flutes (figure 8).

In the very corners, the bearing will bump into the lap shoulder, leaving an area where B is not traced down. A few minutes with bandsaw and file, or chisel, or spindle sander, cleans this up for a perfect fit.

Though procedures like these may become convoluted, care will always yield a perfect fit. Sometimes, of course, I choose to work at risk, not to seek the tedious but foolproof solution. But at other times, when my ultimate concern is the accuracy and strength of the finished joint, I use techniques like the ones I've described here, ignoring the automatic censor that won't allow things of great difficulty to suggest themselves. New joints can emerge from design processes that are integral with and complementary to the lines of the piece as a whole. Such an exposed, routed joint can be the whole focus of an otherwise simple piece of furniture. □

Jim Sweeney makes furniture in San Francisco.

Relying on the Router
Three tricks from San Rafael

Unlike woodworkers in the European tradition and, to a lesser degree, those on the East Coast, many woodworkers on the West Coast came to their profession with a minimum of formal training. Thus there has developed among us a heavy reliance on the electric router for doing quite a few jobs that the old school would have done with other tools in other ways. Never mind that a sharp chisel could do a job well, we're more fluent with the router. A lot of us have spent the better part of a day—or even a week—perfecting a jig or a template that will harness the router to the task at hand.

Often this router technology is enormously efficient, allowing us to repeat processes quickly and accurately. At other times the router work is mundane, just another way of doing an ordinary woodworking job that could be done by other tools. But not surprisingly, the router is also capable of making many cuts and shapes that no other tool can accomplish.

Those of us who rely on the router in our everyday work are the ones who are likely to discover the creative things it can do. Many times it seems that the router jig itself is our final product, rather than the piece of furniture. Sometimes we miss out on the sound of a sharp blade slicing into the wood; instead we put on ear protectors to mute the whine of the three-horse Stanley. Nevertheless, the router technology we've evolved allows an economical efficiency that the professional woodworker cannot overlook in his struggle to make a living. I share shop space with two other craftsmen, Dale Holub and Bruce McQuilkin. The following pages demonstrate what I mean about the router by explaining three of the things that have evolved here in San Rafael: an inlaid wooden handle that Holub routs into his drawer fronts, the quick mortising method that I use, and a nifty wooden hinge that McQuilkin uses for desks and cabinets.   —*Grif Okie*

Holub's inlaid wooden drawer pull

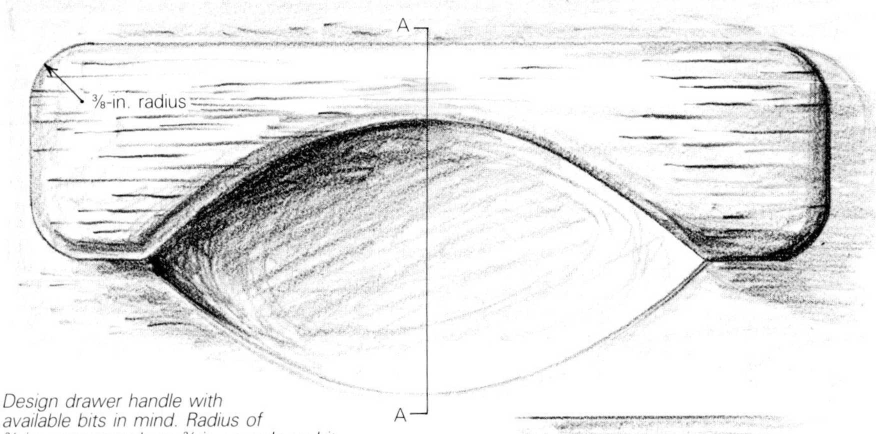

⅜-in. radius

A

A

A-A

Handle insert is undercut for finger clearance Face sits proud of drawer front.

⅜-in. radius

Rout drawer front ⅝ in deep.

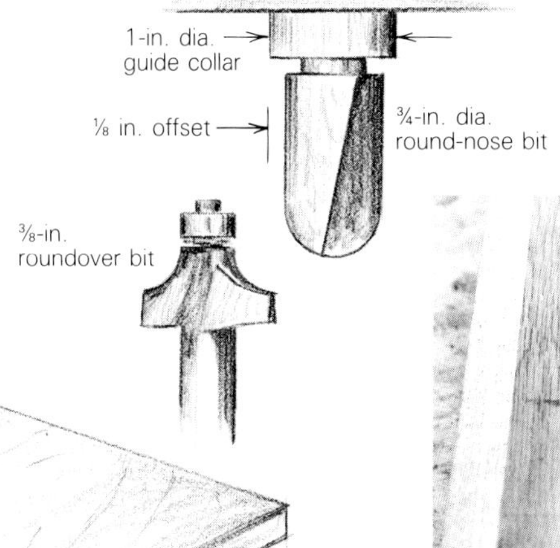

Design drawer handle with available bits in mind. Radius of ⅜ in. corresponds to ⅜-in. roundover bit, and ¾-in. round-nose bit. Drawer front is routed out to accept handle blank and to provide finger clearance.

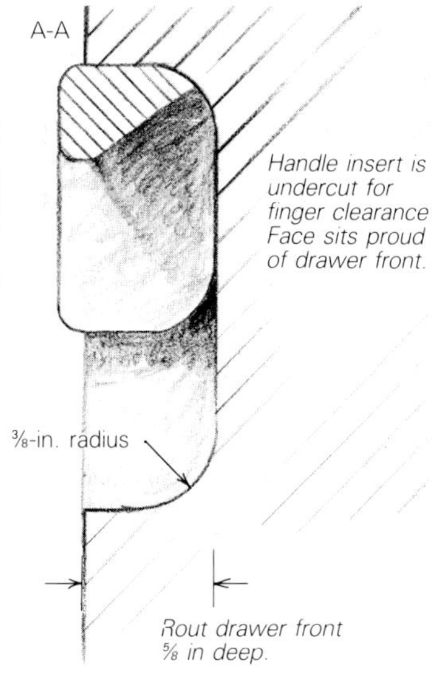

1-in. dia. guide collar

⅛ in. offset

¾-in. dia. round-nose bit

⅜-in. roundover bit

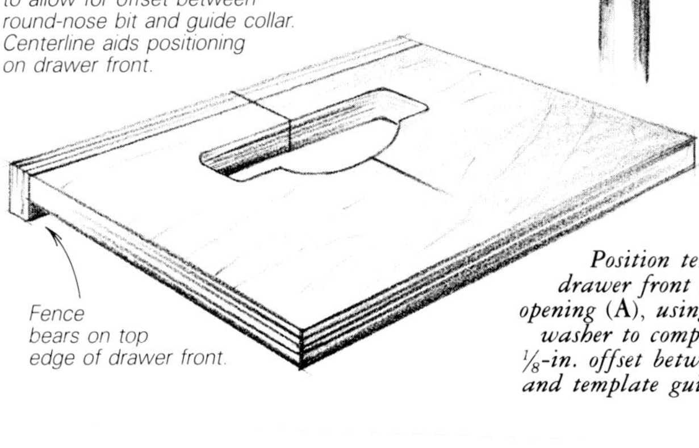

Cut out template for drawer front, adding ⅛ in. all around to allow for offset between round-nose bit and guide collar. Centerline aids positioning on drawer front.

Fence bears on top edge of drawer front.

Position template on drawer front and trace opening (A), using suitable washer to compensate for ⅛-in. offset between cutter and template guide collar.

A

Drawings and photos (except where noted): D. Fillion

B

C

Drill the outlined area ½ in. deep (B). In general, waste as much stock as possible with other tools before routing to the line. Clamp template to drawer front. With a ¾-in. round-nose or corebox bit set to cut ⅝ in. deep, begin routing in a circular motion from the center of the opening (C). With a clear area in the middle, check the depth of cut, adjust if necessary, then rout out the entire area. Clean up with a gouge, and sand the top edge.

D

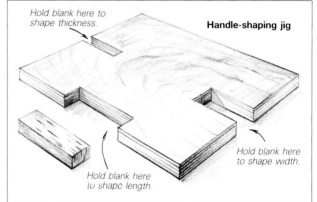

Hold blank here to shape thickness.

Handle-shaping jig

Hold blank here to shape length.

Hold blank here to shape width.

E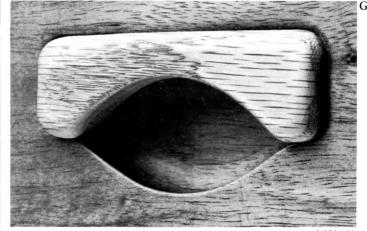

Plane, rip and crosscut handle stock to size. Shape the underside of the blank on the router table, using a ⅜-in. roundover bit with guide bearing (D), and a jig for holding the work (drawing). Alternatively, if you are making a number of handles, you can rip and shape a suitable length of stock to width and thickness, then crosscut individual blanks to length for further shaping. A routed handle blank is shown at left (E). Clean up edges with a drum sander, and try the fit. Transfer the curve of the finger allowance to the face of the handle. Bandsaw the curve of the handle and undercut the finger space (F). Sand the edges. Chamfer or round over the show edges, then glue the handle in place (G). The finished drawer pull is oak inlaid into a koa drawer front.

F

G

Joel Schopplein

Okie's quick jig for routing mortises

This jig for routing mortises is simple to build, and it produces an accurate joint that's easily repeated. It does, however, make a round-cornered mortise. The corners can be squared off with a chisel, or the tenon can be rounded over to match. But you can also rout equal mortises in both pieces of wood and then insert a separate, spline-type tenon. Tenons can be ripped by the running foot, then routed with a roundover bit of the appropriate radius to fit into the round-cornered mortise.

The jig consists of two shoulder blocks clamped onto the wood, with four pieces of hardboard tacked onto them to confine the travel of the router's template collar. The template can be a rectangular hole cut into a single piece of hardboard; I find it easier, however, to make each jig anew to fit the thickness of the stock I'm using, by assembling the jig right on the stock. First decide on the length, width and depth of the mortise you want, according to the strength needed as well as the bits and template guide collars you have. The best mortising bits are either spiral endmills or straight, two-flute cutters, carbide or steel. Lay out the mortise on the stock, then clamp the shoulder blocks to

Okie's mortising jig, for use with router and guide collar, is expeditious. First lay out mortise, then clamp shoulder blocks to faces of stock, and tack hardboard template parts to blocks. Size of opening depends on offset between cutter and guide collar.

the adjacent faces. Make sure that the shoulder blocks are wide enough, of sufficient length, and flush with the surface to be mortised. Now tack or glue the hardboard strips to the shoulder blocks, far enough away from the layout lines to account for the difference in radius between your router bit and its guide collar. With the jig in place, tip the router into the cut and run its guide collar around the inside of the opening you've made, proceeding in suitable increments to the full depth.

Two things make this type of cut easier: a template opening that's wider than the guide collar, and template

hardboard that's thick enough to contact and align the guide collar before the bit gets into the wood. A plunge router (see page 45) is ideal. Walk the router clockwise around the mortise, in the bit's direction of rotation.

I use this method enough to have settled on the mortise widths and tenon thicknesses appropriate for my work and my tooling. For each width of mortise, I make up a bunch of hardboard template end-plates with tongues, as shown. Then making a new jig for each new mortising situation is only a matter of cutting the side plates to length, and nailing them to shoulder blocks.

McQuilkin's inlaid wooden cabinet hinge

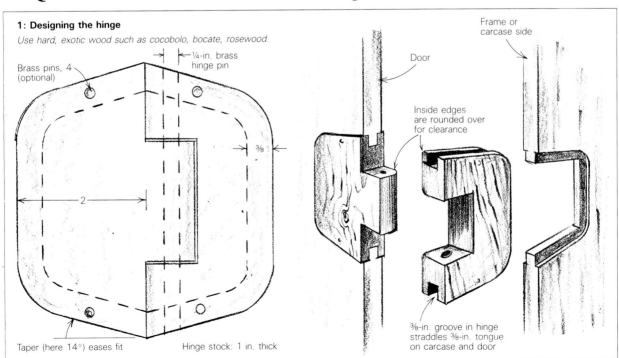

1: Designing the hinge
Use hard, exotic wood such as cocobolo, bocate, rosewood.

Brass pins, 4 (optional)

¼-in. brass hinge pin

2

⅜

Taper (here 14°) eases fit

Hinge stock: 1 in. thick

Frame or carcase side

Door

Inside edges are rounded over for clearance

⅜-in. groove in hinge straddles ⅜-in. tongue on carcase and door

2: Routing the opening

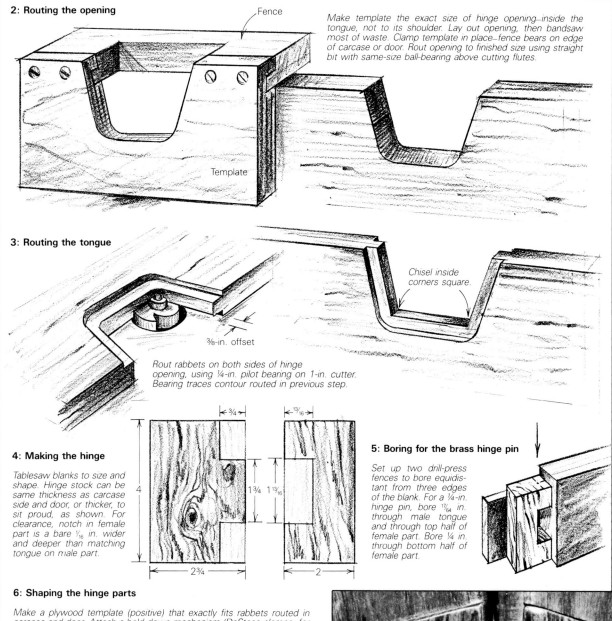

Make template the exact size of hinge opening—inside the tongue, not to its shoulder. Lay out opening, then bandsaw most of waste. Clamp template in place—fence bears on edge of carcase or door. Rout opening to finished size using straight bit with same-size ball-bearing above cutting flutes.

Fence

Template

3: Routing the tongue

Chisel inside corners square.

⅜-in. offset

Rout rabbets on both sides of hinge opening, using ¼-in. pilot bearing on 1-in. cutter. Bearing traces contour routed in previous step.

4: Making the hinge

Tablesaw blanks to size and shape. Hinge stock can be same thickness as carcase side and door, or thicker, to sit proud, as shown. For clearance, notch in female part is a bare ¹⁄₁₆ in. wider and deeper than matching tongue on male part.

¾   ¹³⁄₁₆

4   1¾   1¹³⁄₁₆

2¾   2

5: Boring for the brass hinge pin

Set up two drill-press fences to bore equidistant from three edges of the blank. For a ¼-in. hinge pin, bore ¹⁷⁄₆₄ in. through male tongue and through top half of female part. Bore ¼ in. through bottom half of female part.

6: Shaping the hinge parts

Make a plywood template (positive) that exactly fits rabbets routed in carcase and door. Attach a hold-down mechanism (DeStaco clamps, for example) to position the hinge blanks for shaping on the router table, with 1⅛-in. flush-trim bit and top-mounted guide bearing. Bearing runs on plywood template.

7: Final shaping

Tablesaw ⅜-in. groove around hinge blocks. For safety, dado cutter should tightly fit slot in saw's throatplate. Use ⅜-in. roundover bit to shape inside edges of hinge and carcase for clearance. Glue hinge in place. The finished hinge at right is bocate, let into a rosewood door.

Curved Slot-Mortise and Tenon
Contoured joinery for enhancing frames

by Ben Davies

Design decisions in woodworking cannot be made entirely for aesthetic reasons. Wood is not a plastic medium but a rigid one, and we usually shape it by removing portions of it. Our designs are thus limited by the capabilities of our cutting tools and our skill at using them. To achieve new shapes—to experiment with line and form and the basic geometry of joining wood—we either develop specialized tools or adapt our old tools to perform in innovative ways. Highly specialized hand tools, like molding planes, have limited applications, while more versatile modern tools, like computer-controlled carving machines, can be afforded only by industry. So there's considerable reward for the craftsman in being able to extend the use of general-purpose tools—the router in particular—in imaginative ways.

The following description of making a curved joint is not meant to be definitive. Rather, it is a tentative first step toward adding a dimension to our work when struggling to achieve a balance between geometric and organic forms. When we build we are faced with a dichotomy—crisp and differentiated forms on the one hand, soft and flowing forms on the other. Consider the rigid control exemplified by Shaker and Cubist formalism contrasted with the flowing asymmetry of Art Nouveau. The dichotomy transcends woodworking and the visual arts. For more, read Nietzsche's discussion of the Apollonian/Dionysian duality in his essay "The Birth of Tragedy from the Spirit of Music."

Using a router equipped with an ordinary straight-face bit and a pair of guide bushings, plus a shop-built fixture to hold the work and a bearing template to guide the router cut, you can quickly contour the adjoining shoulders of rails and stiles with little chance for error. But making the fixture and template requires careful planning and accurate work.

Preparing the stock

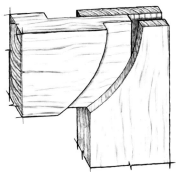

Rail

Stile

Facing shoulders of joint are contoured; rear shoulders remain square. This compensates for the loss of mechanical strength that comes from reducing the gluing surface on the front cheek of the tenon. To prepare the stock, dimension frame members and cut to length. Slot the ends of the stiles as though making an ordinary slip joint; then cut a tenon cheek on the rear face of each rail, but don't remove any stock from the front faces, as these will be routed to produce the curved shoulder shown.

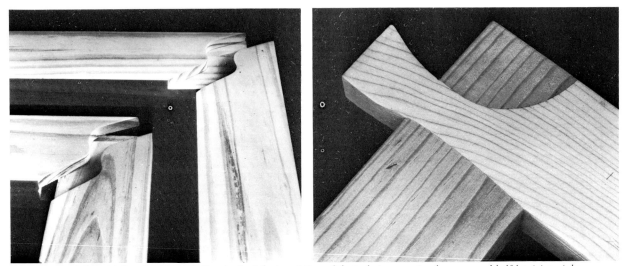

Reverse curve defines joining shoulders of rail and stile, left. A variation of the technique can produce a curved half-lap joint, right.

Photos: Charles Turner

The door on this cabinet shows how contoured joinery can be used to advantage. The wood has been carefully selected so the grain conforms to the curvature of the joint, which repeats the gentle curve made by the bottom sides of the cabinet. The two bottom joints of the door frame have been cut square, complementing the upper corners of the case.

Making the fixture—This part of the system consists of a plywood base and four rabbeted cleats as in the drawing at left. Its job is to hold the rails and stiles and to support the bearing template. The base should be made from a piece of ¾-in. plywood about 15 in. to 24 in. square, a suitable size for joining the frames of cabinet doors. The cleats should be cut from stock whose thickness equals the thickness of the frames plus the thickness of the bearing template, usually ½ in. If you're joining ¾-in. thick frame members, the cleats must be 1¼ in. square, rabbeted to an exact depth of ½ in. and to a width of about ½ in.

To set up the fixture, position the cleats, rabbets in and up, on the edges of the square base; use a true framing square to orient the cleats at precisely 90° to one another (other angles are possible), and screw them to the base with countersunk wood screws. The cleats should not meet at the corners; you have to space them far enough apart so your stock will slide easily through the gap.

The guide bushings—Most routers are designed to accept standard guide bushings generally available as accessories. With a ½-in. bit, use a ⅝-in. O.D. bushing with your router, but any bushing of this general size will do. Because the bushing bears against the curved template when making a cut, and because you're cutting complementary curves using the same bearing template, the line of the cut must be offset from the curvature of the template, and two bushings are required—a large-diameter one for making the cut on the rail,

Fixture Plan view showing bearing template screwed in place

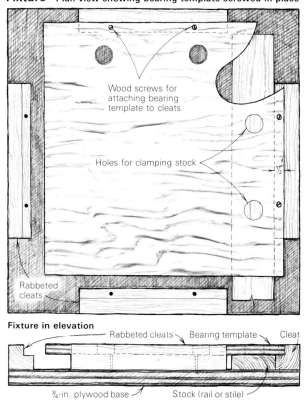

Wood screws for attaching bearing template to cleats

Holes for clamping stock

Rabbeted cleats

Fixture in elevation

Rabbeted cleats — Bearing template — Cleat

¾-in. plywood base — Stock (rail or stile)

and a small-diameter one for cutting the stile (as is the case with the examples here).

Purchase two ⅝-in. guides from the manufacturer; one will serve permanently to hold the outer bushing (epoxied to it), the other as the inner guide bushing. The outer bushing should be turned from brass or aluminum. It is necessary to observe the following mathematical relationship between the diameter of the cutter (D_c), the O.D. of the inner guide bushing (D_{ib}) and the O.D. of the outer guide bushing (D_{ob}):
$$D_{ob} = D_{ib} + 2(D_c).$$

The bearing template—This step involves making three separate templates: one that exactly duplicates the curved line of the joint, another whose profile is offset from this curve and parallel to it, which serves as a pattern for the third tem-

Guide bushings

Section through router base and bushings

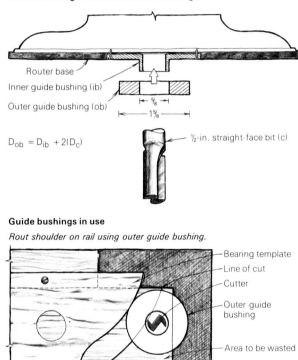

Router base
Inner guide bushing (ib)
Outer guide bushing (ob)
⅝
1⅝

$D_{ob} = D_{ib} + 2(D_c)$

½-in. straight-face bit (c)

Guide bushings in use

Rout shoulder on rail using outer guide bushing.

Bearing template
Line of cut
Cutter
Outer guide bushing
Area to be wasted
Cleat
Entry area

Rout shoulder on stile using inner guide bushing.

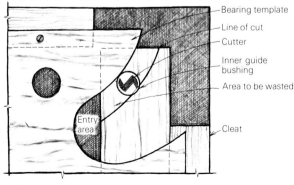

Bearing template
Line of cut
Cutter
Inner guide bushing
Area to be wasted
Entry area
Cleat

Making the bearing template

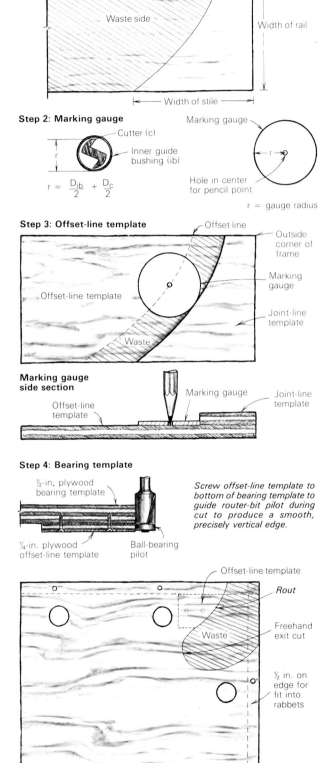

Step 1: Joint-line template

Joint line
Outside corner of frame
Waste side
Width of rail
Width of stile

Step 2: Marking gauge

Cutter (c)
Inner guide bushing (ib)
$r = \dfrac{D_{ib}}{2} + \dfrac{D_c}{2}$

Marking gauge
r
Hole in center for pencil point
r = gauge radius

Step 3: Offset-line template

Offset line
Outside corner of frame
Offset-line template
Marking gauge
Joint-line template
Waste

Marking gauge side section

Marking gauge
Joint-line template
Offset-line template

Step 4: Bearing template

½-in. plywood bearing template
¼-in. plywood offset-line template
Ball-bearing pilot

Screw offset-line template to bottom of bearing template to guide router-bit pilot during cut to produce a smooth, precisely vertical edge.

Offset-line template
Rout
Waste
Freehand exit cut
½ in. on edge for fit into rabbets

plate, the bearing template itself. After composing the curve of the joint on paper, taking into account the width of the rails and stiles, transfer the line to a piece of ¼-in. plywood; then cut along this line with a band saw or jigsaw and smooth the contoured edge with a file. This becomes the joint-line template (step 1 in the drawing at left).

The next thing to do is to make a marking gauge that will allow you to scribe a line on a second template parallel to the curve of the joint-line template. The gauge is made from a plastic disc with a hole in the center for a pencil point, as in step 2. The distance from the outside of the gauge to the pencil point equals the distance from the cutting circle of the router bit to the opposite outside edge of the inner guide bushing (r), which also defines the smallest possible radius of curvature in the joint.

Cut along this offset line and smooth the sawn edge with a file. For the bearing template itself, dimension a piece of ½-in. plywood so that it is slightly smaller than the inner dimensions of your fixture and so its corners are absolutely square. On two adjacent sides, scribe lines that are exactly ½ in. in from the edge and parallel to it. These lines mark the boundary defined by the inner edges of the cleats on the bottom of the bearing template. Now position the offset-line template as shown in step 3 and screw it to the bottom of the bearing template. It should abut one of the cleat lines and be positioned out from the joint line (drawn on the bearing template) with the help of the marking gauge. Using a straight-face flush cutter with its pilot bearing against the offset-line template, cut the curve in the bearing template (step 4). It may seem like a lot of trouble to make one template just to cut another, but only by routing the curve on the bearing template can you get walls that are smooth and perpendicular to both faces.

Next bore four 1¼-in. diameter holes in the bearing template; these permit you to clamp the rails and stiles into place for routing. Also bore four pilot holes for wood screws on the template's edges and countersink them on both sides—top and bottom. This completes work on the fixture and bearing template.

Routing the joints—After preparing the stock, place the frame members in their respective positions and mark each end as shown in the drawing at right so you can orient them properly in the fixture. With the bearing template screwed into position, clamp one of the stile ends (S1) into place so it is flush against the inside cleat and so the end-grain edge is lined up with the inside edge of the top cleat. Attach the inner guide bushing to the router, and set the bit to a depth that equals the thickness of the cheeks of the mortise on the stiles. Insert the bit and bushing into the entry area and rout towards the corner, holding the bushing firmly against the bearing template. Make only a single, careful pass when routing the shoulders on the stiles. Now rout the opposite end of the other stile (S3).

Remove the inner bushing and attach the outer bushing. Insert one of the rail ends (R1) into the fixture so it is flush against the top cleat and its end-grain edge is in line with the inside edge of the right-hand cleat. Rout the tenon on the rail, first wasting most of the stock and finally making one decisive pass with the bushing pressed firmly against the curve of the bearing template. Now rout the tenon on the opposite end of the other rail (R3).

Positioning of stock for routing sequence

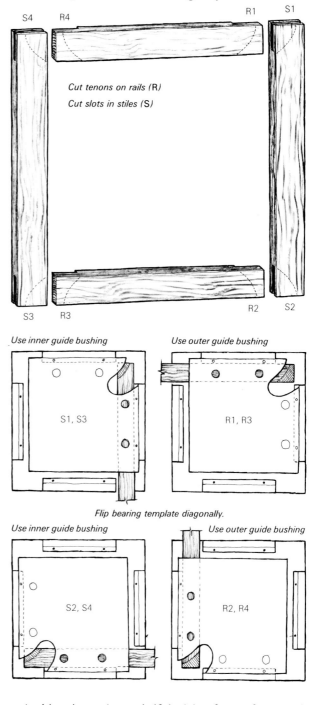

Cut tenons on rails (R)
Cut slots in stiles (S)

Use inner guide bushing

Use outer guide bushing

S1, S3

R1, R3

Flip bearing template diagonally.

Use inner guide bushing

Use outer guide bushing

S2, S4

R2, R4

At this point you've cut half the joints for one frame, and they are diagonally opposed to one another. To cut the other pair of joints, you will have to flip the bearing template and screw it in place on the opposite corner of the fixture. Then repeat the entire process described above, now for S2, S4, R2, R4. On the back side of the frame the joints are left square, which increases the strength of the joint. When you're done, all four joints should fit snugly, and their curves should match up without a flaw. □

Ben Davies owns and operates Muntin Woodworks in Chattanooga, Tenn. His article on frame-and-panel entry doors appears on page 56.

Decorative Joinery
Leading the eye around the piece

by John E. N. Bairstow

The most important element in the craftsman's repertoire is the wooden joint. Although its functional development has been extensive, fascinating possibilities remain unexplored for using the joint in a decorative capacity. Historically, decoration has more often been supplementary, applied to the piece of furniture, rather than integral with its construction. Carving, inlay and marquetry have been used extensively in various forms, while joinery, although potentially the most interesting element, remains quiescent. Few historical examples exist where the method of construction plays a major visual role in the finished work. Thus the designer has become accustomed to creating an attractive piece of furniture using shaped and decorated parts, while sticking to standard joinery beneath.

I choose to start designing a piece of furniture by considering the decorative possibilities of its joinery. This approach makes it possible to use fairly simple forms, and to create the initial visual impact through the joint. Most standard carcase joints, the dovetail and the finger joint, for instance, rarely relate well to the form of a piece of furniture because they concentrate all the interest along the corners. This is fine if you are close to the piece and can appreciate the proportion and accuracy of the joints, but if you are viewing it from more than a few feet, then the most striking thing will be the overall form and not the beauty of the detail. By designing decorative joints that extend beyond the locality of each corner, it is possible to stimulate that first impression at the joint yet bring the viewer's eye around the rest of the piece.

Each of the joints discussed on the following pages is an elaboration of a basic joint, in most cases the finger joint. I do not set out to design a decorative joint with a particular machine or process in mind, but I do try to produce them all with equipment usually available to the designer/craftsman. Small-workshop machines, particularly the router, are versatile, and we should look to them to help carry out creative processes rather than sticking to their conventional functions.

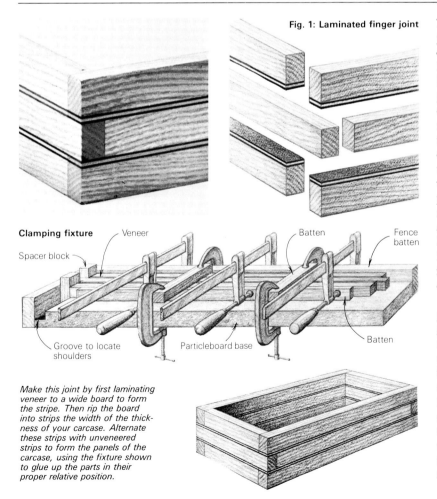

Fig. 1: Laminated finger joint

Clamping fixture — Veneer — Batten — Fence batten
Spacer block — Groove to locate shoulders — Particleboard base — Batten

Make this joint by first laminating veneer to a wide board to form the stripe. Then rip the board into strips the width of the thickness of your carcase. Alternate these strips with unveneered strips to form the panels of the carcase, using the fixture shown to glue up the parts in their proper relative position.

The first joint, illustrated at left, is probably the simplest both visually and constructionally. To make it, begin by laminating veneer to the face of a wide board to produce the stripe in the panel. The thickness of this board, including the veneer combination, will be the width of the laminated fingers in the joint and is therefore an important dimension. When the glue is dry, rip the board into strips, the widths of each will be the thickness of the panel. Each panel (case side) is made from a combination of these strips and of unveneered strips. They are crosscut to alternate lengths; the shorter ones determine the inside dimension of each side of the finished carcase, and the longer ones determine the outside dimension.

When all the strips have been prepared, a simple fixture is required to glue up the panel while maintaining the staggered formation at the ends. It may be difficult to get each strip to lie flat with only side pressure of the clamps. I have used a press that provides both vertical and horizontal pressure, although it is possible to surface and thickness-plane these panels after glue-up without affecting the joint, provided that the grain of all the strips runs in the same direction.

Photos: John Bairstow

Make this joint by first shearing off the corners of the components on a guillotine miter box, shaper or table saw. Then slot the components with a router, a shaper or a table saw, and as-

Component proportions

Finger

Finger is half as wide as panel component

2x

Panel component

semble using plywood splines. A veneer strip can be added between the panels, as shown in the drawing at right, to visually link the corners of the carcase.

The second example, illustrated above, is again basically a finger joint, but its construction is entirely different from the previous one. Visually, the joint is confined to the corner, which is something I try to avoid. This drawback can be overcome by gluing up the panels with veneer between the components, so that thin lines connect the V-shaped "fingers" from corner to corner.

Cut your stock to the required length, width and thickness. The length is the distance between shoulders of the constructed panels (the inside dimensions of the carcase). The width of each com-

ponent is twice that of the finger in the joint. If you want to use contrasting lines to visually link the corners, apply them to the edges at this stage. The thickness of these veneers should be included in the width of the components. Next, cut off the corners of each component at 45°. To achieve the accuracy this joint requires, I recommend using a guillotine miter box and making a pattern to clamp to the top of each piece against which you can locate the guillotine blade. Alternately, you can clamp all the components together face to face and remove the corners in one operation

on the spindle shaper or tilting-arbor table saw. The contrasting fingers can also be cut in one of these ways. The panels are glued up with the aid of a fixture like that shown on the previous page for constructing the first joint. With the panels assembled, groove their ends, as well as the end of each finger, for the plywood spline. You can cut the grooves using a router, shaper or table saw; in the latter two cases, cut the groove before gluing the outer components to the panel, and finish by hand. A jig is required to hold the fingers while carrying out this operation.

The third joint, shown at right, employs dovetail pins to lock a miter joint in both directions. You can vary the length and width of these pins for decorative effect. Prepare each panel to the correct length, width and thickness and construct the carcase using a simple miter at each corner. Depending on the size of the carcase, the dovetail slots are cut using either a router table (or shaper) with a dovetail cutter, sliding the carcase over it, or a portable router in conjunction with a simple jig to guide the router over the carcase.

After you have cut the slots, make sliding dovetail pins to fit into them by cutting one angle on the jointer and then cutting to rough width at the opposing angle on the table saw. I next secure each pin to a block and run it through the shaper, though the pin could as well be hand-planed to final fit. The slots can end square or be rounded off. If you prefer the latter, you can round the pins to match on the disc sander.

This joint is basically a miter into which you rout dovetail slots to fit dovetail keys. The size of the keys can vary. The photo, lower left, shows this same feature used in a finger joint.

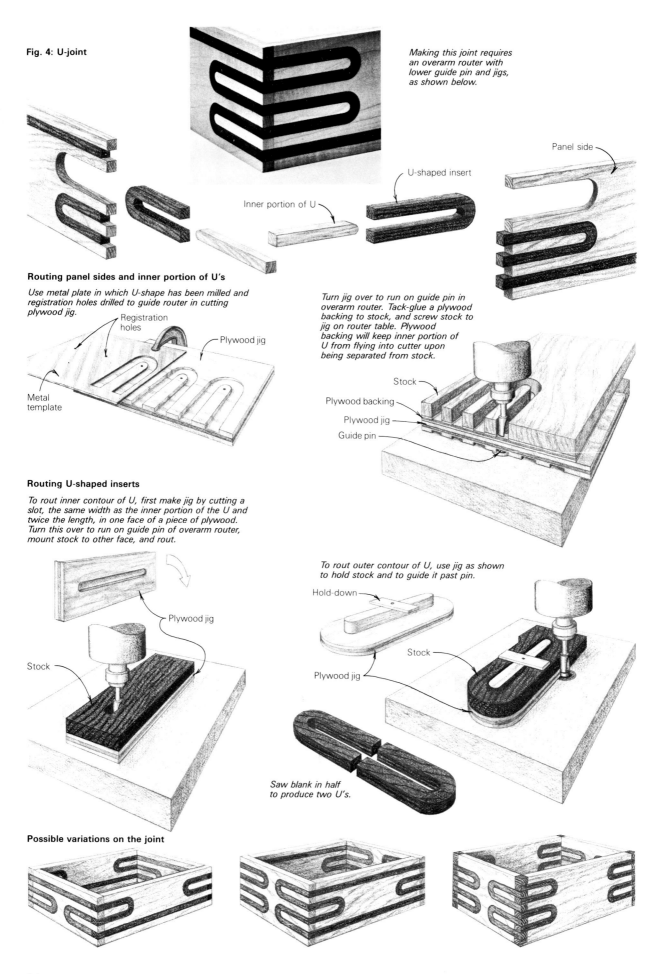

Fig. 4: U-joint

Making this joint requires an overarm router with lower guide pin and jigs, as shown below.

Panel side

U-shaped insert

Inner portion of U

Routing panel sides and inner portion of U's

Use metal plate in which U-shape has been milled and registration holes drilled to guide router in cutting plywood jig.

Registration holes

Plywood jig

Metal template

Turn jig over to run on guide pin in overarm router. Tack-glue a plywood backing to stock, and screw stock to jig on router table. Plywood backing will keep inner portion of U from flying into cutter upon being separated from stock.

Stock

Plywood backing

Plywood jig

Guide pin

Routing U-shaped inserts

To rout inner contour of U, first make jig by cutting a slot, the same width as the inner portion of the U and twice the length, in one face of a piece of plywood. Turn this over to run on guide pin of overarm router, mount stock to other face, and rout.

Plywood jig

Stock

To rout outer contour of U, use jig as shown to hold stock and to guide it past pin.

Hold-down

Stock

Plywood jig

Saw blank in half to produce two U's.

Possible variations on the joint

Probably the most successful of all my decorative finger joints is shown on the facing page. It uses contrasting inserts to create a pattern that flows around the corner, continually leading the eye to and away from the joint. It is possible to link all the joints of a carcase by taking the line the length of the panel. The joint requires the overarm router with the aid of a jig constructed using a metal template. On the milling machine, cut one U-shape into a metal plate and drill the centers for the radii of the remaining U-shapes. From this, rout the plywood jig with all the necessary U-shapes. To retain the inner part of the U, which would otherwise fly into the cutter upon being separated from the stock, glue a thin plywood backing to the stock before routing. Allow for this thickness when setting the cutter depth. Secure the stock to the reverse side of the jig, and place the jig over the guide pin in the overarm router table. Make the initial cuts with a cutter smaller in diameter than the guide pin, to waste most of the stock. After routing, the plywood backing can be pried off the stock and the inner portion of the U's retained. To make the U-shaped inserts, rout the inner contour in an oversized blank (large enough to accommodate two U's, which will later be sawn apart), then use this negative space to locate the blank on a jig that guides the router around the outside contour.

When all the components have been made, it remains only to glue them back into the voids in the panels. The inserts are of such length that the joint is created in this operation. Because the contrasting U also forms a finger in the joint, visual continuity is interrupted by the end grain. This can be overcome by mitering each insert, leaving the end grain of the light wood only. This reduces the gluing area of the joint but not enough to jeopardize its strength.

Fig. 5: Segmented finger joint

Make this joint of any number of finger/panel components, angling the ends to form the pattern. Tack-glue the lengths together before gluing up the panels, which can be done in a jig similar to the one shown in figure 1.

The final joint, above and right, is the most time-consuming to produce, but the many variations possible make the effort worthwhile. The length and cutoff angle of the fingers can create any pattern the designer wishes. Basically, each panel is made from a number of strips with the alternate ones creating the pattern. Each strip is milled to the dimension of the finger in the joint. The decorative pieces are then cut to length and the required angle cut on a guillotine miter box. In this example, each consecutive strip varies by an angle of 15° to produce the curved effect. The contrasting pieces are tacked together with glue, end to end, which holds them until the panels are constructed. The end of each plain strip forms the shoulder of the joint, so the length of the strips should be finished to the inside dimension of the carcase. Each panel can then be glued up as for the joint shown in figure 1, page 82.

These are just five of the many joint variations I have designed—all give a decorative effect when used on anything from small boxes to the largest of cabinets or tables. There are many other ways to use joints decoratively, and no matter how bizarre any idea may seem, it might be quite effective put to proper use. I always make a sample of any joint I design to assess its visual effect, experience the problems it will give in production and decide how it can be efficiently made. Of course, these joints take longer to produce than conventional ones, but the advantage is that any decoration needed in a piece of furniture is already built in. □

John Bairstow designs and builds furniture in Loughborough, England.

Three Decorative Joints

Emphasize the outlines with contrasting veneers and splines

by Tage Frid

I've been a craftsman and designer for 53 years and a teacher for more than 30, but I'm still learning. My students keep me on the ball by always asking questions. I experiment to come up with new ideas and simpler or better ways to do things. Students usually don't ask for help until they are in trouble. By then they have a big investment in time and materials, and we have to figure out some way to fix the mistake so it does not stick out like a sore thumb.

Dovetails are difficult for the beginner, and I have many times shown how to fix a badly fitting dovetail by inserting a piece of veneer. When I thought more about this trick, I realized you could outline the whole joint with veneer of a different color for a nice decorative effect. The technique also works on other joints, such as the mortise-and-tenon slipjoint. Another kind of decorative joint is a three-way miter where the strengthening splines are also emphasized in a contrasting wood. This is an attractive joint for framed cabinets, tables and stools. Here is how to make these three joints.

Outlined dovetails—The joint is laid out, cut and fit in the same way as a regular through dovetail. The veneer inlay that will outline the base of the pins and tails is glued onto the inside face of the mating pieces before the joint is cut. The rest of the outlining is done after the joint is glued together. Gauge the usual depth-lines around the ends of both pieces. To house the inlay, cut shallow rabbets up to the gauge line on the inside face of each piece. If you cut the rabbet on the tablesaw, set the blade as high as the gauge line (the thickness of the dovetailed pieces). Then set the fence to cut the rabbet slightly shallower than the thickness of the veneer.

The grain of the veneers should run in the same direction as the grain in the pieces to be dovetailed. It's easier to trim the inlay flush after the glue has set than to fit it perfectly before. So cut the veneer slightly oversize. Be sure the joint is perfectly tight where the end grain of the veneer meets the solid wood, especially on the pins piece because the veneer will be visible on both edges. Glue and tape the inlay in place

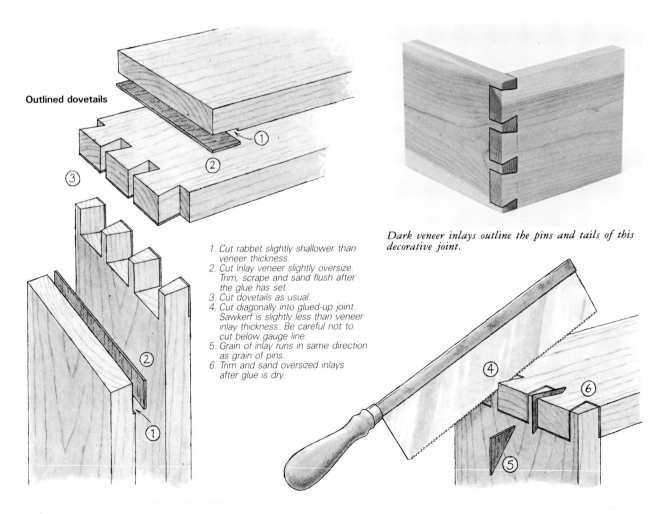

Outlined dovetails

1. Cut rabbet slightly shallower than veneer thickness.
2. Cut inlay veneer slightly oversize. Trim, scrape and sand flush after the glue has set.
3. Cut dovetails as usual.
4. Cut diagonally into glued-up joint. Sawkerf is slightly less than veneer inlay thickness. Be careful not to cut below gauge line.
5. Grain of inlay runs in same direction as grain of pins.
6. Trim and sand oversized inlays after glue is dry.

Dark veneer inlays outline the pins and tails of this decorative joint.

Drawings: Roland Wolf

and clamp it tight. When the glue has dried, lightly scrape and sand the inlay flush.

Now cut and glue up the dovetails. The veneer will line the base of the pins and tails. To add the veneer that will complete the outlining of the joints, saw diagonally along the line of the joint between tails and pins. Use a saw that cuts a little thinner than the thickness of the veneer inlays, and be sure the sawcut doesn't go below the gauge lines. Cut triangular pieces of veneer for the inlays. Orient the cuts so that when the pieces are glued in, their grain will run in the same direction as that of the pins. To fit the inlay pieces in the thinner sawkerf, you need to compress them a little by hammering them or by squeezing them in a steel vise.

Now put some glue in the kerf—not on the veneer. Rub it into the sawcut, using your finger to force it in deep. Slide the veneer into the kerf. It will pick up moisture from the glue and swell for a perfect fit. When all the inlays have been inserted and the glue has dried, cut off the veneer with a sharp chisel and finish-sand.

Outlined mortise-and-tenon slipjoint—This joint can also be decorated with inlay, in the same way as dovetails are. Before you cut the joint, rabbet the inside edges of each piece for veneer. If you cut the rabbet on the tablesaw, use a backing block for more bearing surface against the fence. Flush off the glued-on veneer, then cut and glue up the joint as usual. To complete the veneer outline, saw diagonally down the line between tenon and mortise. Cut veneer triangles slightly larg-

er than finished size, and compress them to fit the kerf. Rub in glue as for the dovetail; you can use a mechanics' feeler gauge to get the glue all the way in. Slide the veneer in, trim it and finish-sand the joint when the glue is dry.

A surer, easier way to make this joint is to glue the veneers on the two cheeks of the tenon before the joint is put together. Rabbet and veneer the inside edges of the two pieces as before. Then cut the tenon and glue the veneer onto its cheeks. Allow for the veneer thickness when laying out the tenon thickness. Cut the mortise to fit the veneered tenon and glue up the joint as usual. For dovetailed or slipjointed pieces made of thicker wood, the inlay could be thicker too.

Decorative splined miters—There are other ways besides a veneer outline to emphasize joints. A strong, decorative and quite simple joint to make is the splined miter frame, as shown on the next page. I made this three-way miter frame joint with wood that is square in section. For demonstration, I made only one corner joint—in a table you might have four, in a cabinet, eight (one at each corner of a cube). Glue together the joints of the mitered frames. I use hot hide glue because it sets fast. Next, cut the grooves for the decorative splines. If you cut the grooves for the splines on a tablesaw, use a cradle to hold the piece at a 45° angle to the table. I cut a notch in a 2x4 to make a cradle. For strength and decoration I put in several splines.

Clean up the surfaces after the glue has dried, then bevel or miter one side of each frame along its length so they will

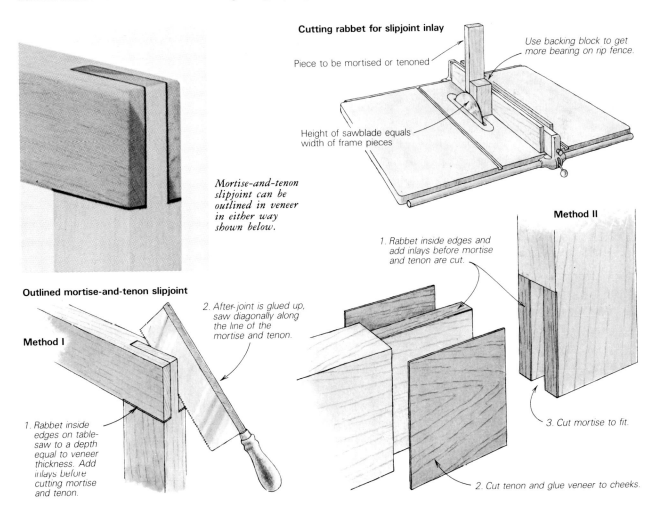

Cutting rabbet for slipjoint inlay

Piece to be mortised or tenoned

Use backing block to get more bearing on rip fence.

Height of sawblade equals width of frame pieces

Mortise-and-tenon slipjoint can be outlined in veneer in either way shown below.

Outlined mortise-and-tenon slipjoint

Method I

1. Rabbet inside edges on tablesaw to a depth equal to veneer thickness. Add inlays before cutting mortise and tenon.

2. After joint is glued up, saw diagonally along the line of the mortise and tenon.

Method II

1. Rabbet inside edges and add inlays before mortise and tenon are cut.

2. Cut tenon and glue veneer to cheeks.

3. Cut mortise to fit.

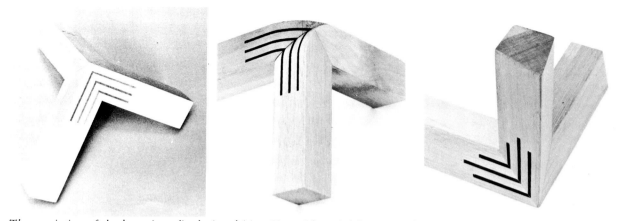

Three variations of the decorative splined mitered joint. Piece with angled faces, at right, is the most difficult of the three to make.

Decorative splined miter

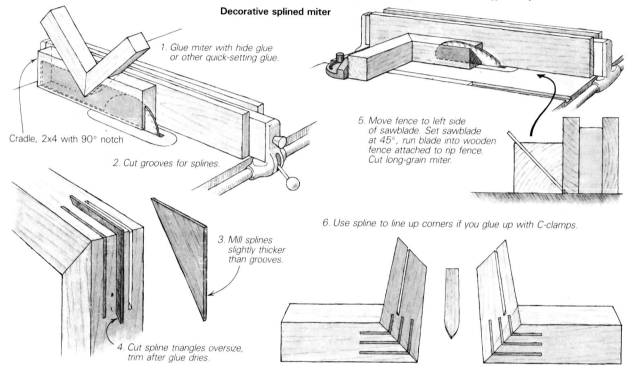

1. Glue miter with hide glue or other quick-setting glue.

Cradle, 2x4 with 90° notch

2. Cut grooves for splines.

3. Mill splines slightly thicker than grooves.

4. Cut spline triangles oversize, trim after glue dries.

5. Move fence to left side of sawblade. Set sawblade at 45°, run blade into wooden fence attached to rip fence. Cut long-grain miter.

6. Use spline to line up corners if you glue up with C-clamps.

fit together. You can do this with a hand plane or with the tablesaw. With the tablesaw, mount a piece of wood on the rip fence, tilt the blade to 45° and run the blade slightly into the wooden fence. Use trial and error to find the right setting of blade and fence. You can leave a $\frac{1}{32}$-in. shoulder on the mitered piece, to bear against the fence beyond the sawcut. Plane this shoulder off before gluing up, or lose it later by rounding the edge. Don't stand directly behind the blade when you make this cut—the waste can be thrown backward.

Next make a groove in the mitered side for a hidden or blind spline, using the tablesaw or an electric router. The joint is long-grain to long-grain, so this spline is not for strength but for getting the corners to align if you glue up using clamps. If instead you wrap strips of inner-tube around the joined pieces, stretching it as you go, the corners will align and the spline won't be necessary. I have also used $\frac{1}{2}$-in. surgical tube; it's inexpensive and works better.

You can round off the corners, as in the photo, top center, before gluing up. I shaped the curves on a disc sander and dry-fit them to make sure they lined up at the joint.

For an interesting effect, you can put an angle of about 15° on the faces of the frames, as in the photo, top right. This joint is more difficult to make. Before gluing the frames together I ripped one face of each piece at a 15° angle. These bevels all should be on the front faces (in the same plane). You can use the same kind of tablesaw setup as for ripping the miters. These cuts must be very accurate. Finish-sand these faces before mitering and gluing up each frame. Then cut the sawkerfs and glue in the splines.

Next miter a long-grain side of each frame and cut the groove for the positioning spline if you are gluing up with clamps. Then rip the outside faces of each frame (that is, where the decorative splines appear) at 15° and finish-sand them. Finally, glue the two frames to each other. I like this joint and am going to use it in a frame-and-panel cabinet—when I get the time. □

Tage Frid, a Danish cabinetmaker, is professor emeritus of woodworking and furniture design at the Rhode Island School of Design in Providence.

Template Dovetails
Another way to skin the cat

by Charles Riordan

Tools for template dovetails.

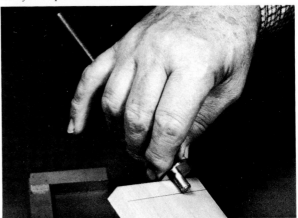

For centuries the dovetail joint has been the most satisfactory method of joining two pieces of wood meeting at the end grain. It's especially appropriate for drawer and case construction. The joint is strong, durable and decorative, one of the hallmarks of the master craftsman. However, the tyro need not approach dovetailing with a faint heart. There are indeed many ways to divest a feline of its fur; I believe using templates is the best way to produce dovetails. The method described in this article, with special attention to drawer construction, owes its inspiration to Andy Marlow, whose *Classic Furniture Projects* (New York: Stein and Day, 1977) describes a similar approach.

Templates can be made quickly and easily using the sheet aluminum stocked by almost any hardware store. The photo top right shows the tools I use. The scratch gauge, photo right, is a machinist's gauge made by Starrett Tools (Athol, Mass. 01331), and I use it for both laying out the templates and marking the boards for the dovetails. Over the years I have found this marking tool easier to use and more accurate than the conventional cabinetmaker's marking gauge.

Charles Riordan makes period furniture in Dansville, N.Y.

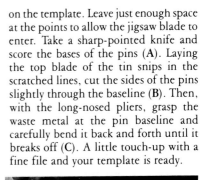
Machinist's scratch gauge marks wood and metal.

To make the template, cut a piece of aluminum about 4 in. wide and long enough to span the width of the stock you are joining. Make sure that the bottom is square to both sides. Don't count on the sheet you buy at the hardware store being square at each corner; check it. Set the scratch gauge to the thickness of your drawer-side stock, or the stock you will cut tails in, and mark the sides of the template. You will be able to make two templates from this one

sheet, one on each long side.

Determine the number of pins necessary for the depth of the drawer and mark their centers on the template using either a rule at an angle or the dividers, starting so the bottom pin just misses the groove for the drawer bottom. I use the bottom of each board as my reference edge for marking the pieces. With the proper pin spacing marked, set the bevel to the desired angle, usually 15°, and mark the pins

on the template. Leave just enough space at the points to allow the jigsaw blade to enter. Take a sharp-pointed knife and score the bases of the pins (**A**). Laying the top blade of the tin snips in the scratched lines, cut the sides of the pins slightly through the baseline (**B**). Then, with the long-nosed pliers, grasp the waste metal at the pin baseline and carefully bend it back and forth until it breaks off (**C**). A little touch-up with a fine file and your template is ready.

A. *Scoring the template baseline.*

B. *Snipping the template.*

C. *Breaking out the waste.*

Hold the boards to be marked for the tails firmly on your bench either by clamping or with bench dogs. Leave enough space above the dog to hold the square against the bottom and end of the board, and butt the template against the square (D). Hold the template firmly and mark the tails (E). It's important here that the scratch awl be sharpened to a needle point and that it be held at the same angle for each marking, an angle that will ensure that the point follows the template cutout unerringly. Observe the grain direction and mark so that the grain will force the point of the scratch awl against the guide. The consistency of the marking will determine the accuracy of the joint.

To mark the pins, I place the board to be marked in the vise with a flat back-up board extending slightly above it. If your vise is too narrow to exert pressure across the whole width of the board, use a C-clamp or two to make sure there is no gap between the back-up board and the piece to be marked. Place the template on the end of the board to be marked and firmly against the back-up board. Abut the template against a stop held against the lower edge of the piece to be marked (F) and mark the pins. Here again I must stress the importance of the angle at which the scratch awl is held; a few degrees variation will result in a poorly fitting joint, especially if the errors are additive.

With a small square and the scratch awl, mark the inside face of the drawer back (or front) for the wide side of the pins. I find that it helps also to mark the narrow side of the pins on the outside face for through dovetails, especially if the wood is not fairly straight-grained. Now take a soft, black pencil that has been sharpened to a chisel point and trace lightly over the scratch marks (G). This will leave the center of each scratch line clearly defined as a thin, light line between the black pencil lines.

To cut the pins for a lipped (rabbeted) drawer front I use a router, with the drawer front clamped vertically in the vise (H). To give the router a firm base I clamp a piece of hardwood 3 in. thick, 4½ in. wide and 12 in. long to the face side of the drawer front. I also clamp a stout piece of hardwood to the inside of the drawer front both to protect the piece from clamp marks and to act as a stiffener. Instead of using the router guide fence I use a fence (or back stop) fastened to the heavy base board. I find that this gives firmer control over the router and less chance for tipping or wobble when routing end grain. This is

D. *Positioning template on tail board.*

E. *Marking the tail board.*

F. *Positioning template on pin board.*

G. *Squaring down the pin lines.*

H. *Routing waste for half-blind dovetails.*

important because the router bit is extended to include the depth of the drawer lip as well as the pins and is otherwise unstable.

Do not attempt to rout all the way to the pin lines. Come as close as you feel is safe, leaving a shaving or two to clean off later with the chisel. Of course, the depth of the pins and their width is taken care of by the depth of the router bit setting as well as by the positioning of the fence.

The next step is to cut the tails and rough out the pins of the drawer back (I) on the jigsaw. A band saw can be used, but a jigsaw will give better results. Use a sharp, fine-toothed blade, 3/32 in. to 1/8 in. wide, tensioned enough not to wander. Don't depend on your unmagnified eyesight here. I use 4× magnifying glasses and seldom have to touch a chisel to the tails or pin bases after jigsawing. Saw just to the waste side of the scratch-awl markings on the inside face highlighted by the pencil.

After the pins are cut on the jigsaw they will, of course, be square pegs that must be brought to their final triangular shape with a paring chisel. Don't try to chop down in one or two large bites. Nibble off fine shavings as shown (J). The chisel must be as sharp as a razor. The first cut will give the indication of which direction the grain is going. If the grain runs into the pin, pare in horizontally from one face or the other, holding a stout piece of wood against the opposite face of the pin so the wood will not break out as the chisel cuts through. Here again, use the magnifiers and place the chisel in that fine, light line between the pencil markings to make the final shave.

To cut away the shavings and clean up around the base of the pins, I use a spear-pointed chisel I made from a 1/4-in. straight chisel by grinding it to a 30° point (K). It is much better than a skew because it cuts on both sides and can be held flat on the pin baseline while severing the shavings. Its usefulness is fully appreciated when cleaning up the pins for a blind dovetail joint, as no square-edged chisel can quite get into the acute corners.

Now that you have carefully followed that fine, light line with sawblade and chisel and a careful inspection shows no more trimming to be necessary, you can assemble the pieces (L), and tap the tails home using a hammer and a block of wood. If you have done all your cutting with the care needed to make a dovetail joint, the pieces should go together with a firm "thunk" and you'll have to go out and buy a new hat because your old one will never again fit.

Glue up only after you're sure all the parts fit with no binding or excessive pressure being brought to bear. Apply glue to the pins only and clamp with just enough pressure to snug up the joints. Too much will cause bowing and will produce an out-of-square drawer or case with poor joints. Use a try square on the inside corners to make sure this is not taking place. □

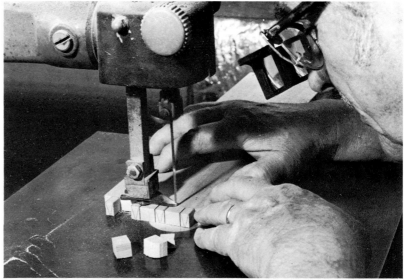

I. *Sawing waste from through dovetail pins. Do the tails the same way.*

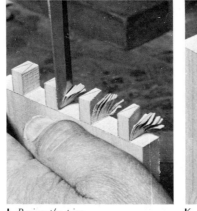

J. *Paring the pins.*

K. *Cleaning pin bases with spear-pointed chisel.*

L. *The completed joint.*

Gluing Up
How to get a strong, square assembly

by Ian J. Kirby

Gluing up is unique among woodworking operations. It's an irreversible, one-shot deal and has to be got right. You may have done accurate work up to this point, only to find that a small error in assembling or clamping has produced all sorts of inaccuracies that will be difficult, perhaps impossible, to fix. A common lament in woodworking is that "everything went perfectly until glue-up, then everything went wrong." When you think about it, this is not surprising. How often do we systematically consider gluing up, and how much time do we give to dry clamping? Usually very little, and then halfheartedly.

To get the best results, we should bring a studied method to this operation and practice it more. We ought to have a table especially prepared for this purpose, its top surface flat, clear of debris and well waxed to resist glue penetration. A piece of varnished plywood over your benchtop will do, but a sturdy table, 36 in. high with a Formica surface, is better.

Before gluing up, you should dry-clamp each assembly exactly as you would clamp a glued assembly. This means positioning and tightening all the clamps, with correct glue blocks, and checking the whole assembly for accuracy. Gluing should proceed calmly, in an atmosphere of preparedness, with the glue and necessary applicators ready, clamps standing by, and you and your assistant decided on the order of events. The time of day you glue up is important. Most woodworkers like to glue up in the evening and let the glue set overnight. To meet this goal, a lot of work often gets rushed, dry clamping gets short-circuited and we have all the necessary ingredients for a disastrous glue-up—fatigue, unpreparedness and anxiety. The only reason to proceed under such conditions lies in the spurious notion that glue cures only while the moon is out.

Consider the alternative. Leave the work dry-clamped overnight. The next morning, check the clamping to see if everything is still properly aligned. Then collect all the tools and materials you need and begin to glue up. The light is better, your mind is fresh, the pressure to complete the job is gone. If you can't leave the work dry-clamped overnight, at least let it sit for an hour while you attend to other things.

Gluing up actually begins with a decision about what to glue together and in what order. The more subassemblies you can get together, the easier the total operation will be, especially the final glue-up. But before gluing any parts, always clean and prefinish surfaces that cannot be reached later with a plane. It's much easier to work on a piece of wood while its entire surface is exposed and accessible than to try to remove mill marks and other blemishes once other parts are permanently in the way.

You must also weigh the inconvenience of having three or four finishing sessions as the job proceeds against one grand and difficult cleanup after final assembly. Prefinished surfaces will resist glue penetration from squeeze-out. When the

excess glue is dry, simply lift it off with a sharp chisel. When finishing prior to gluing up, take care to keep the finishing material off joint interfaces.

Bar clamps—The type of bar clamp you use has considerable bearing on the ease and accuracy of the glue-up. For general cabinetmaking applications, quick-action bar clamps with a circular pad at the screw end are no substitute for a standard bar clamp. A good clamp should sit on a glue table without falling over at the slightest touch. The bars should be identical in section, and the heads should move easily, but not flop around. When pressure is applied, the face of the head should be at 90° to the bar—this way we know exactly where pressure is transferred. A collection of clamps is an investment no matter which you choose, but it is a long-term investment you can make by purchasing one or two clamps at a time. The combined value in the end may equal that of two major machines. The best choice in the main is between Jorgenson, Wetzler and Record clamps. I prefer the last.

Applying the glue—The amount of glue squeeze-out is an important signal. Since it is waste, it is best to have as little as possible, but we still want assurance that there is sufficient glue in the joint. The smallest bit of squeeze-out is enough. This results from getting the glue in the right place in the right amount. For different jobs you will need different applicators. White glue (polyvinyl acetate) and yellow glue (aliphatic resin) can be stored in and dispensed from a squeeze bottle. But a squeeze bottle is not a good applicator, and won't guarantee that the surfaces being joined are completely wetted by the glue. There is no reliable adhesion if joining surfaces don't get completely wetted.

Since the future of the piece depends on the quality of its joints, we need to take a close look at the business of applying the glue. Manufacturers try to lower the surface tension of the glue so it will spread easily. Nevertheless, the glue should be rubbed or rolled on, not merely squeezed out onto the surface. A set of stiff-bristle brushes of different sizes (I use plumber's flux brushes) will suffice in most instances. If you use white or yellow glue, brushes can be stored in a jar of water, but it is just as easy to wash them out and begin next time with a dry brush. If you use urea-formaldehyde glue or resorcinol, you must clean the brush after each use, for these will set hard even in water.

When edge joining boards, white or yellow glue can be squeezed from a plastic bottle onto one surface. If the boards are clamped or rubbed together and the clamps are removed immediately and the joint broken, chances are you will find that the glue has covered both surfaces uniformly. That this happens in edge joining does not mean that it will happen in other joining situations. In fact, it's not a good method for edge joining either. A better way would be to run a very light

The size and position of the clamping block can make the difference between success and failure when gluing up. The block should be the same dimensions as the section of the rail, and placed directly opposite the shoulder area, centering the clamping pressure on the joint.

bead on each surface, and then with a 1-in. wide paint roller, spread the glue thinly. Now we know that the surfaces are evenly wetted and that when the joint is clamped we won't have gobs of the stuff dripping on the floor, the table and the clamps. Spreading glue with your fingers is a bad practice. You need fingers for other things, and the grease and dirt you add to the glue won't help adhesion. If you insist on using your fingers, wash your hands first.

Edge gluing—When gluing up several narrow boards to make a case side, tabletop or framed panel, there are four important considerations: the position of each board in the composite piece, the grain direction of each, the number of clamps to be used and the means for aligning and registering the boards to keep them from swimming about when pressure is applied. Assuming that all of the boards are dimensionally stable and free of defects, the first thing to decide is how to arrange them to get the best appearance. This involves shuffling the pieces around to achieve visual harmony and continuity in the figure. Remember that after gluing up you will have to clean up the surfaces with a smoothing plane, so try to decide on an arrangement that permits the grain of all the boards to run in the same direction. If this is not possible, then you will have to plane in both directions and carefully avoid tearing the grain on an adjacent board.

Having decided on their arrangement, mark the boards so their order can be recalled when they are put in the clamps. Dry clamping will determine the number of clamps you need and where to put them. As shown in figure 1 on the next page, clamping pressure is diffused in a fan of about 90° from the clamp head. You will need enough clamps to ensure that the lines of diffusion overlap at the first edge joint. The number of clamps then is a function of the length of the boards and of the widths of the two outer boards. Since the boards themselves transmit and spread the pressure, clamping blocks are unnecessary in edge gluing. Plan to joint and rip the composite piece to width after glue-up. This will remove any depressions the clamps leave.

When you know how many clamps the job calls for, put half the number (half plus one if the total number is odd) on the table, evenly spaced. If the bars are not bent or damaged, they will register the boards in the horizontal plane. Coat each edge to be joined with glue, wetting all the surfaces thoroughly and uniformly. When the boards have been put

into the clamps and slight pressure applied, place the remaining clamps over the top of the panel and begin to tighten all the clamps. Having an equal number of clamps top and bottom prevents the panel from bowing under pressure.

Ideally you should be able to edge-glue boards without having to rely on any mechanical means of holding them flush. But when the boards are even slightly warped this isn't possible. It is common to use dowels for aligning and registering boards; if you're going to do this, use a doweling jig to align the holes. Another method for registering edge-glued boards is the Lamello joining plate. The machine that cuts the slots is worth the investment for the professional woodworker —or you can make your own (see page 67). The quickest solution is to lightly clamp battens across the width of the panel. Don't overtighten the bar clamps. This can squeeze out most of the glue and starve the joint. Moderate pressure is all that is needed when edge gluing.

Gluing up mortise-and-tenon joints—Mortise-and-tenon joints usually get little attention when gluing—most woodworkers want to assemble them as fast as they can. But the pace ought to be less hurried; ideally there should be two people at a glue-up, one to direct the order of events and to tighten the clamps, the other to manage the shoe end of the clamps. Both apply the glue. One coats the tenons thoroughly while the other puts glue in the mortises. Because the mortise and tenon goes together in a sliding fit you can't expect to apply glue to the tenon alone and still have enough in the joint. You will have to spread glue in the mortise as well. Don't just squirt in some glue and push it around with a stick or pencil, because the excess glue can impede fitting the joint. Visualize the glue as a fluid pad of considerable thickness. The pressure exerted by an excess on one side of a tenon can misalign the members. So apply glue thinly and evenly to all surfaces. Better than squeezing lots of glue into the mortise and stirring it with a stick is using a stiff-bristle brush.

During dry clamping pay close attention to dimensioning and positioning the clamping blocks. Their purpose is not so much to protect the stock as to transfer the pressure from the clamp to the workpiece in exactly the area required. The fact that the clamp heads can lean away from a right angle under pressure, that they may have been put onto the work slightly askew and that the workpiece may not have edges perpendicular to its face, are all things you must consider when direct-

Fig. 1: Edge gluing

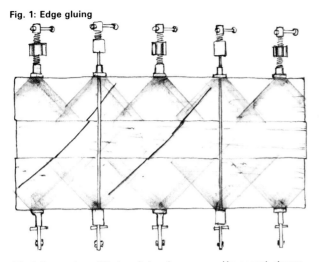

Shaded areas show diffusion of clamping pressure. Use enough clamps to ensure that lines of pressure overlap at the two outer joints. An equal number of clamps above and below the panel prevents bowing.

Fig. 2a: Gluing leg/rail assemblies — Plan view

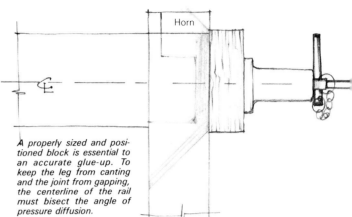

Horn

A properly sized and positioned block is essential to an accurate glue-up. To keep the leg from canting and the joint from gapping, the centerline of the rail must bisect the angle of pressure diffusion.

Position the blocks initially by eye, but check across the inside surface of both legs with a straightedge to make fine adjustments in their final placement.

Fig. 2b: Side elevation

Correct   Incorrect

Leg   Rail   Straightedge

Clamping block   Misaligned block

Fig. 3: Checking for alignment

Use a single pinch rod to measure the diagonal from an inside corner to the bottom of the opposite leg. If diagonals are precisely the same, the assembly is square.

Make pinch rods from stable, straight-grained wood. You can glue a block to the tip of each rod before cutting the points. This will hold the rods above obstructions in the assembly. In use, extend the points into opposite corners of a frame, pinch the rods together and then check the other diagonal.

Fig. 4: Gluing-up flat frames

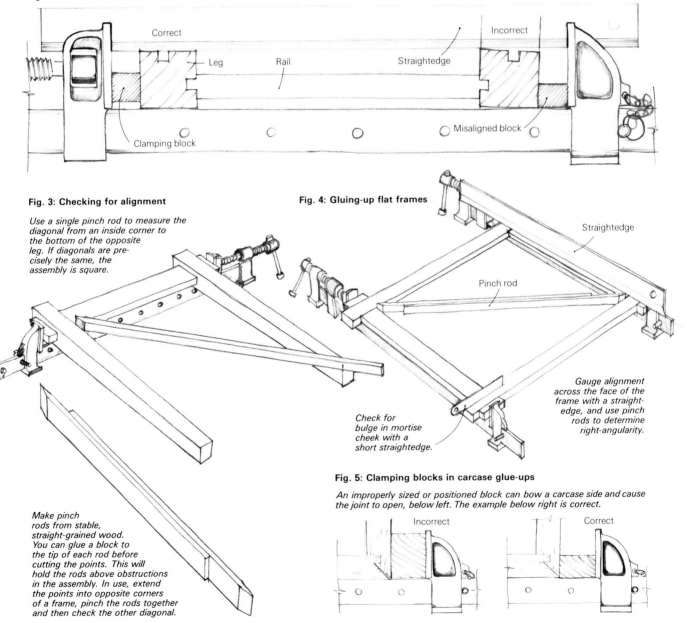

Straightedge

Pinch rod

Check for bulge in mortise cheek with a short straightedge.

Gauge alignment across the face of the frame with a straightedge, and use pinch rods to determine right-angularity.

Fig. 5: Clamping blocks in carcase glue-ups

An improperly sized or positioned block can bow a carcase side and cause the joint to open, below left. The example below right is correct.

Incorrect   Correct

ing clamping pressure. Putting a piece of plywood between the workpiece and the clamp shoe or head isn't enough.

In gluing up leg/rail or rail/stile assemblies, the size of the glue block that distributes the pressure to the shoulder line of the joint is important. Providing that the grain of the block runs lengthwise and it is thick enough not to distort under pressure, the block should be about as long as the rail is wide, and about as wide as the rail is thick, as shown in figures 2a and 2b. With the right clamping block, pressure can be placed where you need it by moving the block slightly to one side or another or up or down the legs. If you attempt to glue up without clamping blocks, there's little chance of directing pressure where it is required.

The parts of a correctly glued-up assembly or subassembly should not twist or wind in relation to one another. They should be aligned and at right angles to one another. Joint lines should close up tightly. When assembling two legs and a rail, as in a table frame, cut the legs ¾ in. longer than the finished length. This excess, called the horn, is left at the top of the leg, where it can reinforce the mortise and help keep the end grain from splitting during dry clamping and gluing up. The horns are cut off later when the glue has cured. On rail/stile assemblies, where you have mortises at both ends of the vertical members, add 1½ in. to the length of the stile, making a ¾-in. horn at each end. When laying out the joints on legs, measure from the bottom to the top, not from top to bottom. This way you won't have to cut the legs to final length after assembly. When laying out the joints on stiles, clamp the two members side by side and lay out the final cut to length at the top (striking across both at once); then measure down from these to lay out the finished length at the bottom. Use a try square and a layout knife for the best results. Then you can accurately check the work when gluing up.

Checking for alignment—The legs should be sighted with winding strips to make sure they are in the same plane. Don't try to sight tapered legs on the tapered side. To correct twist or wind in the assembly, one person holds it down tightly on the clamp bed and slackens off the clamping pressure. The second person, holding one leg in each hand, moves the legs into proper alignment. Then the clamps are retightened.

Right-angularity between legs and rail is frequently overlooked. This is best checked by laying a straightedge across both legs as shown in figure 2b. To correct misalignment, the clamping block must be raised or lowered to redirect clamping pressure. If this isn't done, the rails will go off at odd angles at the next glue-up when the subassemblies are joined by two more rails. Then the finished piece will be under constant tension and the rails may bow. Using the straightedge as a reference, shift the clamping blocks in the appropriate direction. Here dry clamping will tell you ahead of time where to position each block. Remember to dimension each block carefully, as improperly sized blocks are difficult to position and can misalign an accurate joint by putting pressure in the wrong place.

Next check for overall squareness in plan. A try square is hardly adequate for checking this sort of right-angularity. On assemblies with long legs and rails, it can gauge only a fraction of the lengths involved, and if the legs are curved or tapered, or if the rails aren't straight across their bottom edges, a try square won't work at all.

Squareness is best determined by taking diagonal measure-

ments from the top inside corners to the inside bottom corners of the legs. If the diagonals are equal, the assembly is square. You can make these measurements with a long rule, though take care to place the rule at the same depth in the two corners. A flexible metal tape can also be used, but this requires two people for accurate results. One holds the one-inch mark in precise alignment with the corner, while the other pulls the tape taut to measure the distance to the bottom inside of the leg.

Traditionally, diagonal measurements are taken with pinch rods. Shown in figures 3 and 4, these sticks have pointed edges that fit into corners to measure diagonals. They are stepped so they can span obstructions such as center stiles and stretchers. Considering their high degree of accuracy and the small amount of effort needed to make them, there's little reason to use anything else to measure internal diagonal distances. We should find the diagonals equal if the clamp holding the assembly together is in line with the rail member. If they are not equal, we will have to reorient the clamp and the blocks in such a way that the members will creep into square with one another.

Assembling frames—When gluing up a flat frame, as for a door, employ the same checking procedures and assembly methods as you would for a leg/rail assembly. But because a frame is closed on four sides and relatively thin in section, it calls for some special attentions. The flatness of the gluing table is particularly important. If we are using identical clamps and we press the frame down onto the clamp bars, we will get a twisted frame if the table we are working on has a twisted top. Usually the cheeks of a mortise in a frame are fairly thin, and the glue in the joint migrates to the center where it can cause them to bulge outward. Avoid this by clamping across the cheeks with a C-clamp and properly sized blocks.

Gluing up carcases—A situation where improperly dimensioned clamping blocks can be dramatically counterproductive is in gluing up carcases. Too large a block, as shown in figure 5, can misapply the pressure, cause the sides to bow like crazy and open the joint on its outer edge. For clamping case sides tightly against an internal shelf or other member, we have little choice other than to use cambered cauls, or we won't achieve the necessary pressure.

Gluing dovetails is totally different from gluing mortises and tenons. If the dovetail has been made so that the end grain of the pins and tails is below the surface on the adjacent boards, then all that will be required to glue a large carcase together is one clamp and two people. After the glue has been applied and the joint has been put together, the dovetails should be clamped home individually. They will not spring back if they are correctly made, because there is considerable friction in the glue interfaces between the pins and tails. When the glue has been uniformly brushed on the long grain of the pins and tails, each part swells, making the close fit even tighter. Don't delay assembly after applying the glue, and don't try to hammer the parts home, as you quite properly did during the dry test assembly. Clamping is the proper means once the glue is applied, and it is sweet and easy to clamp each tail one after the other and see the glue come squeezing out at the bed of each pin. □

Which Glue Do You Use?

Part one: Chemical types, not brands, make the difference

by George Mustoe

Like the alchemists' attempts to transmute base metal into gold, much human effort has gone into the search for the perfect glue. This goal is probably as unrealistic as the dreams of alchemy, but the inventors' struggles have not been without reward: adhesives manufacturing is a big growth industry in the United States, and per-capita consumption is about 40 lb. per year.

Not surprisingly, "What kind of glue did you use?" is a frequent query heard whenever woodworkers gather. Unfortunately, these exchanges generate some old wives' tales, among them the colorful but incorrect assertion that cyanoacrylate "superglue" is derived from barnacles.

Because wood is a relatively weak construction material, most adhesives produce bonds that are stronger than the surrounding lumber, so claims of extremely high strength are seldom meaningful to the woodworker. Instead, the most important characteristics are setting rate, viscosity, resistance to water, flexibility, color, sandability, and gap-filling properties. As a woodworker who happens also to be a chemist, I've developed a keener than usual interest in the literally hundreds of glues sold today, discovering in the course of my research that only about a dozen kinds are useful for woodworking. Within each category, I've found that different brands will usually perform equally, so the choice for a particular project is best made by understanding the chemical makeup and characteristics of the glues we use.

In this article, I'll cover those glues that are best suited to general woodworking. In a second article, I'll talk about epoxies, hot-melt glues, cyanoacrylates and contact cements, all specialty glues that are usually more expensive, though not always better, than our old standbys.

Protein glues—The natural world abounds with examples of sophisticated adhesives which display impressive tenacity; barnacles and mussels, for example, cement themselves to beach rocks and ship bottoms with a substance that resists prolonged immersion in salt water. Though the chemistry of these natural adhesives is poorly understood, most sticky secretions are combinations of various complex proteins. Thus it is not surprising that early artisans discovered that the best raw materials for glue were protein-rich animal products such as skin, bone and blood.

Today, despite the advent of modern synthetic adhesives, animal-protein glues are still common. They can be divided into three types: hide and bone glue, fish glue, and blood glue. Of the three, hide and bone glues are of the greatest interest to the woodworker. The use of fish glue, which is derived from the water-soluble proteins in fish skins, is limited to industry, mainly for attaching labels to bottles and occasionally as a tack-improving additive to white glue. Blood glues, once developed as water-resistant adhesives for early military aircraft, are made by dispersing beef or pig blood in water, with wood dust, lime or sodium silicate added as thickening. They're most often encountered in vintage plywood, but are practically impossible to buy today and have no significant advantages over readily available synthetics.

Hide and bone glues, on the other hand, are far from obsolete. Besides being widely used in industry for products such as gummed paper tape, sandpaper and bookbindings, hide glue finds plenty of uses in the woodshop. The setting time and spreadability can be varied, and the adhesive cures into a colorless, nontoxic, sandable glueline which can be undone by the application of moist heat—a feature that is attractive to luthiers, for instance, who may need to remove the soundboard of an instrument to repair it. Water also softens hide glue, and some furniture conservators use a 50% vinegar solution to speed the disassembly and repair of antiques.

Hide glue consists of protein derived from collagen, the main ingredient of skin and connective tissue. The glue is prepared by cooking animal hides, hooves and tendons into a protein-rich broth which is then cooled to a gelatinous solid, sliced, dried and ground into a coarse powder. In retail stores, hide glue is commonly sold as a pre-mixed liquid, but it can be bought in powder form, in which case it must be mixed with hot water in a glue pot or double boiler—it loses its strength if you let it boil. Liquid hide glues have two advantages over mix-your-own: you don't need a heated glue pot, and the slow setting rate may be valuable for complicated assemblies. During the heydey of hide glue, it could be bought in 18 grades, each with a different viscosity and setting time. Today, woodcraft suppliers usually offer only a single, high-grade product. Setting time can be slowed by adding more water, but this leads to a slightly weaker bond.

Many other proteins have adhesive properties. Soybean-based glue is used in some interior plywood. Casein or milk glue, which has been detected in medieval picture frames, is made from skim milk, and is used today for laminating interior beams and trusses. This glue is a light-colored powder that must be mixed with cold water and allowed to stand about ten minutes before use. Unlike the other protein glues, casein sets both by evaporation and by chemical reaction, forming calcium caseinate. The resulting neutral-colored bond is highly moisture-resistant but not waterproof. Casein can be used in cool weather and on woods containing up to 15% moisture. It is particularly effective with oily woods such as teak, yew and lemonwood. Powdered casein glue is available from National Casein, 601 W. 80th St., Chicago, Ill. 60620.

Petrochemical resins—Casein glue is sometimes confused with polyvinyl acetate (PVA) white glues. Part of this confusion stems from the milky appearance of white glue and also because dairy-related companies such as Borden, who once marketed casein, now sell PVA glue. Developed during the 1940s, PVA glue is part of a family of synthetic resin glues

Polyvinyl acetates are made by some 40 to 50 companies, and as this photo shows, prices for a 4-oz. bottle vary (1983 prices). Mustoe found that expensive brands aren't necessarily better.

that have largely replaced animal glues in the woodshop.

Derived from petroleum compounds and acetylene gas, white glue consists of minute PVA globules suspended in water. When the glue is spread on wood surfaces, the water evaporates and/or diffuses through the surrounding porous material, and the globules coalesce to form a tough film. Because of its reputation as a cheap hobby cement, white glue is sometimes undervalued as a woodworking adhesive. Actually, its low viscosity, rapid setting time and fair gap-filling qualities make it an excellent choice for general woodworking. It dries into a clear, slightly flexible glueline, and it remains fresh on the shelf almost indefinitely. PVA is nontoxic, making it safe for use around children.

The major disadvantages of white glue are its low resistance to moisture and the gummy, thermoplastic nature of the dry film: it turns rubbery under the heat of sanding and clogs the sandpaper. You can minimize clogging by removing excess glue with a sponge or a damp cloth before the glue sets, or by trimming away hardened glue with a chisel or a scraper. The soft film also causes PVA-glued joints to "creep" out of their original alignment when subjected to continuous stress. While this may allow joints to adjust to seasonal variations in humidity without cracking, it's not a desirable quality if great structural strength is needed.

Be cautious when buying white glue. Competition among the 40 to 50 manufacturers of PVAs has kept the price low, but with the predictable advertising hype. Weldbond, for instance, calls its PVA a "concentrated . . . universal space-age adhesive . . . not similar to any other type of bonding agent being offered." In fact, the adhesive contains a lower percentage of solids than Elmer's and several other brands of white glue. Weldbond's most significant characteristic may well be its relatively high price. In testing white glues, I found only one that yielded inferior results, a generic white craft glue distributed by a local hobby shop. Its adhesive properties compared favorably with leading brands, but the glue reacted with most woods to produce gray or black stains. Chemical analysis revealed that the glue was contaminated with high levels of dissolved iron.

In recent years woodworkers have been attracted to another type of PVA adhesive, aliphatic resin or yellow glue. Actually, the label is a bit of a marketing ploy, since both yellow and white glues are technically aliphatics, which means that they consist of long chains of molecules. Yellow glues have qualities similar to those of white glues, but they contain polymers that speed tack time and improve moisture- and creep-resistance, at the expense of a slower final cure. Yellow glues are

also less thermoplastic, so they won't gum up sandpaper as badly. Borden's Elmer's Carpenter's Wood Glue and Franklin's Titebond are two of the best-selling brands.

Yellow glue may be more difficult to apply because of its thick consistency, but it is also less likely to squeeze out when clamped. The viscosity increases as the glue ages in the container. Manufacturers recommend that the glue be used within 6 to 12 months of purchase, but some workers successfully store it for up to two years by stirring in small amounts of water to reduce the viscosity. Up to about 5% water can be added without affecting bond strength. Freezing can ruin white and yellow glues, both in the bottle and as they cure. Manufacturers add compounds to improve freeze-resistance, but any PVA that seems curdled should be discarded.

Water-resistant glues—Modern industrial processes have been revolutionized by the development of highly water-resistant synthetic resins, beginning in 1872 when the German chemist Adolph von Baeyer (of aspirin fame) discovered that he could produce a solid resin if he reacted phenol with formaldehyde. This basic chemistry forms the foundation of the plastics industry and has given birth to a family of versatile, reliable adhesives. Phenolic resins, because of their cost and heat-curing requirements, are used mostly in industry and for exterior plywood and water-resistant particleboard. But a chemical cousin of the phenolic resins, urea-formaldehyde resin, is cheaper and easier to use, making it an adhesive of choice when water resistance is needed, or when long open time between spreading the glue and clamping up is desirable.

Phenolics and urea-formaldehydes cure not by evaporation, but by cross-linking or polymerizing their molecules into hard films that aren't softened by water. The small-shop woodworker will be most familiar with the type that consists of a light brown powder which must be mixed with water before use. Weldwood and Wilhold manufacture this adhesive, both under the label "plastic resin glue." Another type, Aerolite 306, is sold with a hardening catalyst that speeds curing.

Urea-formaldehydes are good general-purpose wood adhesives, especially for woods of relatively high moisture content. They cure into hard, brittle films which won't clog sandpaper, but, for the same reason, they are poor gap-fillers. The medium brown color when cured blends well with most cabinet woods, although bonding may be inhibited in some oily species such as rosewood and teak. Most urea-formaldehydes aren't recommended for marine use, but they are sufficiently water-resistant to withstand sheltered outdoor applications.

When high strength is not essential, urea-formaldehyde can

Adhesive	Application characteristics	Properties after curing	Recommended uses
Hide glue (hot)* (Behlen Ground Hide Glue, Behlen Pearl Hide Glue)	Fast tack, viscous, min. curing temp. 60°F, moderate gap-filling ability, nontoxic, requires glue pot	Transparent, not water-resistant, can be sanded	Musical instruments, furniture
Hide glue (liquid)* (Franklin Liquid Hide Glue)	Slow-setting, low viscosity, min. curing temp. 70°F, moderate gap-filling ability, nontoxic, may have strong odor	Similar to hot hide glue	Assembly procedures that require slow setting
Casein glue* (National Casein #30, slow cure; National Casein #8580, fast cure)	Glue must stand 10 to 20 minutes after mixing, min. curing temp. 35°F, moderate gap-filling ability, nontoxic	Neutral opaque color, high water-resistance, sands cleanly	Interior structural applications, especially good with oily woods and in cool working temperatures
White glue* (Elmer's White Glue, Franklin Evertite, Weldbond, Wilhold R/C-56)	Cures rapidly, low viscosity, min. curing temp. 60°F, moderate gap-filling ability, nontoxic, almost unlimited shelf life	Transparent, low water-resistance, creeps under load, clogs sandpaper	General woodworking, not recommended for structural or outdoor applications
Yellow glue* (Elmer's Carpenter's Wood Glue, Franklin Titebond)	Fast tack, moderate viscosity increasing with age, min. curing temp. 60°F, moderate gap-filling ability, nontoxic	Nearly transparent, moderate water-resistance, less likely to creep under load than white glue, can be sanded	General woodworking, indoor use only
Urea-formaldehyde glue* (Weldwood Plastic Resin Glue, Wilhold Plastic Resin Glue)	Glue powder must be mixed with water, min. curing temp. 70°F**, poor gap-filling ability, releases formaldehyde vapor, uncured glue is toxic	Medium brown color, high water-resistance, sands cleanly, thick gluelines are brittle and may crack under stress	General woodworking, structural uses indoors or in sheltered outdoor locations, bonding may be inhibited with oily woods
Resorcinol glue (Elmer's Waterproof Glue, U.S. Plywood Resorcinol, Wilhold Resorcinol)	Moderate viscosity, min. curing temp. 70°F**, good gap-filling ability, releases formaldehyde during curing, two-part system must be mixed, uncured glue is toxic	Opaque reddish color, waterproof, withstands most solvents and caustic chemicals, can be sanded	Marine use and outdoor construction

* Water-based adhesive may cause warping of veneer or thin panels.   ** May be rapidly heat-cured at 90°F to 150°F.

be extended by adding up to 60% wheat flour or fine wood dust. The thermosetting nature of urea-formaldehyde glues can be both boon and bane. In a shop cooler than 70°F they will cure poorly or not at all, but at 90°F the mixture's pot-life is only one to two hours. Once the glue has been spread and the pieces have been clamped, curing can be hastened by heating the glueline to between 90°F and 150°F. Urea-formaldehyde's thermosetting qualities make it the most popular adhesive for use with radio-frequency curing apparatus, or dielectric gluing.

One drawback of urea-formaldehyde glues is the emission of formaldehyde gas during and after curing. Besides being a suspected carcinogen, this vapor may irritate the skin and eyes and cause headaches. The problem is liable to be most pronounced in homes constructed with urea-formaldehyde-glued paneling, but it's a good idea to work with this adhesive only in a well-ventilated shop.

The development of urea-formaldehydes marked a milestone on the road to the perfect waterproof glue sought by boatbuilders for centuries. Ironically, completely waterproof adhesives didn't appear until the wooden ship was nearly extinct. Today, resorcinol-formaldehyde glue is the most popular waterproof wood adhesive, with epoxy resin trailing as an expensive second choice. Resorcinol glue was developed during World War II for gluing the plywood used in bombers,

helicopter blades and antimagnetic mine sweepers. Today, it is used to bond marine and exterior plywood, and for laminating outdoor timbers. For the home shop, resorcinol is sold retail as a two-part system consisting of a dark red liquid resin, and a solid powder containing paraformaldehyde and an inert filler (usually powdered nutshells). Two brands are Wilhold and U.S. Plywood, both marketed as waterproof glue.

Resorcinol is fairly costly, and once mixed it must be used within an hour or two. For these reasons, it should be the glue of choice only when a completely waterproof joint is needed. It requires a minimum setting temperature of 70°F, and solidifies within eight hours, though it doesn't reach full bond strength for several days. Acidic hardwoods such as oak may require 100°F to 110°F temperatures for maximum bonding. The final glue film is extremely durable, tolerating boiling water, caustic chemicals and drastic temperature variations. Resorcinol glue is easy to apply and can be cleaned up with a damp rag. Disadvantages include the dark reddish glueline and the release of formaldehyde during curing. □

George Mustoe is a research technician in geochemistry at Western Washington University in Bellingham, and a serious amateur woodworker. Photos by the author.

Why glue joints fail

When wood joints fall apart, as they occasionally do, the glue is automatically suspect. Usually, though, bond failure occurs not because the glue isn't strong enough, but because the wrong adhesive was used, the wood's moisture content was too high or too low, the surface was improperly prepared, or the joint was clamped incorrectly.

The wide range of glues available will meet any woodworker's requirements, but for most indoor woodworking, white and yellow glues are the best choice, except for veneering, where water-free glues such as epoxy or hot-melt sheets will keep the veneer from curling. Powdered resin glues can give erratic results due to sloppy mixing or poor temperature control, but they are excellent when a hard, machinable glueline is required, and for moisture-resistant exterior work.

Too much or too little moisture in the wood is one of the most frustrating causes of glue failure. Consider this example: The center of a 2-in. thick board is liable to contain more moisture than the surface. If the lumber is planed and edge-glued before it reaches equilibrium moisture content, the porous end grain of the wood will dry and shrink faster than the middle, straining or breaking the glueline. To avoid this, stack and sticker your lumber after milling, postponing gluing until it has stabilized. An extra coat of finish on the end grain when your project is done will minimize subsequent stress on the glueline.

Climatic extremes can drive wood to equilibriums that will make gluing troublesome. In the desert Southwest, for instance, moisture content sometimes falls to 4%, which can draw the water out of PVAs before the joints can be assembled. Conversely, the glue won't harden at all in wood much wetter than 12%. In these environments, using adhesives that don't cure entirely by evaporation—urea-formaldehydes and casein glues—will help. Temperature can also be a factor in glue failure. Below 50°F, PVAs come out of solution and cure in chalky, weak gluelines. At high temperatures, say, above 100°F, they are liable to skin over before assembly, which makes a strong bond virtually impossible.

Typically, adhesives bond to only the top layers of wood, so the surface must be smoothly cut, with no torn or partially detached fibers. Providing that it is straight and true, the best surface for edge-gluing is one left by a sharp hand plane. Next best is to use a jointer or even a sharp circular saw, preferably one that leaves indetectable sawmarks. Dull jointers and planers, on the other hand, produce a glazed, burnished surface which swells in contact with glue, encouraging failure. A sanded surface is similarly undesirable because the loose fibers left behind by the abrasive soak up glue but will part readily when the joint is stressed. Wood surfaces oxidize quickly, so try to mill and glue on the same day; machining a fresh surface on lumber that has been stacked for acclimation is advisable.

Mating surfaces should fit snugly without massive clamping pressure, but joints should have enough space to permit a glue film to develop; hammertight tenons or dowels will squeeze out most of the adhesive as they are assembled. If a joint is sloppy, don't rely on your glue's gap-filling qualities to rescue it. Better to recut the joint, or to salvage it with a strategically placed veneer shim.

Deciding how much glue to apply is a dilemma often not solved until it's too late. The ideal glueline is as thin as possible, but without starved spots. Thicker lines are generally weaker because they contain air bubbles or trapped solid particles, as well as internal stresses that develop as the adhesive shrinks during curing. Most glues, particularly PVAs, perform best if they're spread on both surfaces, and the surfaces then placed together and allowed to stand for about 10 minutes before being clamped. This "closed time" gives the adhesive time to penetrate and coalesce before the clamps squeeze it out.

To bond successfully, glues require surprisingly little clamping pressure—10 PSI is plenty, more will just squeeze out the glue, starving the joint. The most common clamping problem is an uneven glueline caused by poorly distributed pressure. Obviously, each job calls for its own setup, but a joint is clamped correctly when the glue squeezes out a bit just as the two parts mate, gap-free. Exert more pressure after that and you risk starving the joint or racking the assembly. For edge-gluing, a good rule of thumb is to space clamps at intervals equal to twice the width of each board. So two 4-in. boards should be clamped every 8 in., with generously dimensioned clamping blocks to spread the pressure and to protect the wood. Before actually gluing, dry-clamp your parts. If a joint won't close, fix what's wrong so you won't be tempted to draw it up later with crushing clamp pressure, introducing stresses that make failure probable.   —G.M.

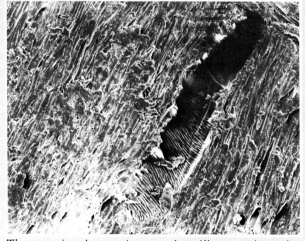

These scanning electron microscope photos illustrate why a crisply cut surface is better for gluing than a handsawn one. The photo of hand-planed maple at left shows cleanly sheared fibers which allow liquid glues to penetrate. The partially detached fibers of a sawn surface, right, limit glue absorption to top layers and break readily under stress.

Glues for Woodworking

Part two: Synthetics solve some problems, pose new ones

by George Mustoe

Adhesives have changed enormously since the days when artisans prepared their own crude glues from meat scraps or milk curds. Twentieth-century chemistry has given us hundreds of new synthetic adhesives, some of which are of interest to the woodworker. Generally these adhesives—epoxies, cyanoacrylates, hot-melt glues and contact cements—are far more expensive than the hide glues, polyvinyl acetates and water-resistant glues that I discussed in the first part of this article, pages 96 to 99, although they aren't necessarily more effective. In the small woodshop, cost alone limits use of most of these glues to special jobs.

Epoxy resins—Although epoxies are among the more expensive adhesives, their physical properties—high strength, low shrinkage, transparency, insolubility, and ability to bond to a diverse array of materials—make them ideal for certain applications. The extreme strength of epoxy is seldom essential in joining wood to wood, but it makes it possible to bond wood to glass or to metal. Cured epoxy machines well, and its dimensional stability makes it an excellent choice for filling gaps and mending holes.

All epoxies are two-part systems: a resin, and a liquid catalyst or hardener. They harden by chemical reaction between the two components, not by solvent evaporation. The glue is activated by mixing the resin and hardener together, usually in equal proportions. Changing the resin/hardener proportion affects the properties of the cured epoxy: slightly increasing the hardener by up to 10% makes the bond more flexible, while increasing the resin by up to 10% makes the bond more brittle. Using a larger proportion of either component weakens the bond.

Curing time for epoxies varies according to temperature. Epoxies generally require temperatures of 65°F or higher to set, although special formulas have been developed for use at lower temperatures. Heating the joint to 100°F to 150°F speeds the setting rate, but also increases the health risk: the vapors from hot epoxy are very toxic. Viscosity drops dramatically at higher temperatures, causing the epoxy to flow out of the joint onto other areas of the work. Whatever the temperature, uncured epoxy is toxic, and repeated skin contact can provoke allergic reactions in some people. Acetone is recommended for cleaning up uncured epoxy, but denatured alcohol works just as well and is less flammable.

Epoxies come in a variety of different types. "Quick-set" or "5-minute" epoxies are convenient where you need a strong, fast-setting bond, but their brief pot life can be frustrating if your assembly takes longer than you expected. They also have less strength and water-resistance than conventional epoxy. Hand-moldable sticks of epoxy putty are easy to mix, and work well as a filler. Opaque "filled epoxies" contain suspended solids such as clay or powdered metal to provide increased strength, higher viscosity or other desired proper-

ties. Filled epoxies have a putty-like consistency that makes them perfect for filling large voids or repairing surface dents, but they have a limited shelf life because the filler eventually settles out of suspension. Vigorous stirring will sometimes restore old stock to a usable condition. When clear epoxy resin gets old, it may become thick and granular (some preparations have recommended shelf lives of only 6 to 12 months), but warming the container to about 100°F in a hot water bath will return the epoxy to its original state.

Many retail brands of clear epoxy are bought in bulk from the manufacturer and repackaged into small containers. There isn't much difference between brands, and you can save money by purchasing epoxy in larger quantities. One ounce of epoxy in a tube costs about $2.25, but Sig Model Airplane Epoxy, an excellent transparent glue found in hobby stores, costs about $7.50 for 12 ounces. Though epoxy manufacturers won't often sell bulk quantities directly to the public, they'll usually provide technical assistance and lists of local distributors. Armstrong Products Company, PO Box 647, Warsaw, Ind. 46580, makes a clear all-purpose epoxy called A-271 resin, a "quick-set" resin called A-36, and several other types. Devcon Corporation, 30 Endicott St., Danvers, Mass. 01923, also distributes epoxy in bulk. As another example of the money you can save by buying in quantity, Devcon 210 epoxy costs $2 an ounce in hardware stores, but about 62¢ an ounce in gallon lots (all 1983 prices).

Several epoxies have been marketed specifically for woodworkers by Industrial Formulators of Canada, Ltd. Their G-1 epoxy is a general-purpose clear resin; G-2 is recommended for oily and acidic woods such as teak and oak. Cold Cure is meant for use at temperatures as low as 35°F, Five Cure sets in 15 minutes or less at temperatures above 40°F, and Sun Cure is a low-viscosity laminating resin. You can mail-order these epoxies from Flounder Bay Boat Lumber, 3rd and "O" Ave., Anacortes, Wash. 98221.

An extensive line of epoxies, additives and dispensing pumps is sold by Gougeon Brothers, Inc., 706 Martin St., Bay City, Mich. 48706, under the trademark West System. Developed for boatbuilding, the Gougeon system uses epoxies both as adhesives and as saturation coatings to prevent transfer of moisture and improve dimensional stability.

Polyester resins—If you want to reinforce wood with fiberglass cloth, your best choice is epoxy resin, but because of epoxy's high cost, polyester resin is commonly used instead. It is also less toxic and much cheaper. Like epoxies, polyester resins are two-part systems: a low-viscosity liquid which hardens when a small amount of catalyst is added. Although polyester resin performs well for reinforcing fiberglass, it lacks sufficient wetting ability to bind to wood fibers. It will adhere to wood only if the solidified resin can interlock with surface irregularities. You can get an adequate bond if you roughen

Adhesive	Application characteristics	Properties after curing	Recommended uses
Epoxy Armstrong A-271, Cold Cure, Devcon 210, Five Cure, Industrial Formulators' G-1 and G-2, Sig Model Airplane Epoxy, Sun Cure, West System	Moderate viscosity, decreasing greatly at warm temperatures, min. curing temp. 65°F**, excellent gap-filling ability, two-part system, liquid and vapors are toxic, may cause skin irritation	Highly transparent, waterproof, bonds to many materials, sands and machines well	Marine and outdoor use, excellent for bonding nonporous materials
Cyanoacrylate Devcon Zip Grip 10, Duro Super Glue, Eastman 910, Elmer's Wonder Bond, Hot Stuff, Krazy Glue, Scotchweld CA-3	Very fast bonding***, very low viscosity, min. curing temp. 60°F, poor gap-filling ability, odor may be irritating, bonding may be inhibited by oil or acidic residues, excess is very difficult to clean up	Highly transparent, very water-resistant	Small repairs, modelmaking, bonding nonporous materials
Hot-melt stick Bostik Thermogrip, Swingline Fix Stix	Sets almost instantly, high viscosity, excellent gap-filling ability, does not penetrate well, difficult to apply over large areas, nontoxic, glue gun needed	Neutral opaque color, moderate strength, remains slightly flexible, cannot be sanded, softens when heated	Furniture repairs, small projects such as toys, construction of jigs and patterns
Hot-melt sheet BF Goodrich Plastilock 810	Instant bonding, applied with hot iron, poor gap-filling ability	Moderate strength, softens when heated	Veneering
Contact cement, solvent-based Weldwood Contact Cement, Wilhold Contact Cement	Instant bonding, min. curing temp. 70°F, poor gap-filling ability, highly toxic, very flammable, difficult to clean up	Low strength, creeps under load	Bonding plastic laminates to plywood, not recommended for veneering, though it is widely used for this purpose
*Contact cement, water-based (latex)** Elmer's Cabinet Maker's Contact Cement, Weldwood Acrylic Latex Contact Cement	Instant bonding, min. curing temp. 65°F, poor gap-filling ability, low toxicity, easy to clean up, "open time" must be carefully monitored according to label directions	Very low strength, water-resistant, creeps under load	Recommended when limited ventilation conditions prevent using solvent-type cement

   * Water-based adhesive may cause warping of veneer or thin panels.
  ** May be rapidly heat-cured at 100°F to 150°F.
 *** Surface activator speeds set.

the surface with a rasp or coarse sandpaper. Brush the catalyst-activated resin over the wood and allow it to soak in thoroughly. Before it begins to set, apply glass cloth and brush another coat of activated resin over the cloth.

Unlike epoxy, polyester resin shrinks considerably after curing, and it may remain slightly tacky long after solidification. To produce a smooth surface finish, polyester finishing resins sometimes contain emulsified wax which floats to the surface as the resin cures.

You may be able to save up to 50% by purchasing polyester resin and fiberglass cloth from a local business that uses them. Boatyards are usually willing to sell a gallon or two of resin from their 55-gallon drums at minimal mark-up. These industrial-grade resins are sometimes slightly red or purple in color, compared to the water-clear retail resins. You can color clear polyester resin by adding pigments specially made for this purpose.

Cyanoacrylate glues—No adhesive has received more attention in the last few years than cyanoacrylate, commonly known as "superglue." Contrary to rumor, cyanoacrylate is not derived from barnacles (a misconception that has stuck as well as the glue). Nor is it new—the first cyanoacrylate adhesive, Eastman 910, was discovered by accident during a test

of the light-refracting properties of a new organic compound when a drop was placed between glass prisms and they stuck fast. The glue was patented and first marketed in 1958, and industry has been using it ever since.

Cyanoacrylate will bond most plastics and rubber, and is good for gluing rubber to wood or to metal. Higher-viscosity formulas are sold for use on wood and other porous materials. Elmer's Wonder Bond Plus and Krazy Glue for Wood and Leather are two brands available in small retail packages. As most users have discovered, cyanoacrylate also has a remarkable ability to bond skin. Glue distributors now sell solvents to dissolve unwanted bonds, although acetone and nail polish remover are somewhat helpful for this purpose.

Cyanoacrylate provides a very water-resistant, but not completely waterproof, bond. Prolonged immersion in water eventually weakens the joint. This adhesive will also resist most organic solvents.

Cyanoacrylate glue is most useful in modelmaking, musical instrument building and other small-scale applications. Its main advantage is its extremely rapid set—3M's Scotchweld CA-3, for example, sets in about 30 seconds. You can reduce this setting time to as little as one second by brushing on 3M's Scotchweld Surface Activator for Cyanoacrylate Adhesives before you apply the glue. Moisture will also speed curing, but on wet surfaces the glue will leave chalky stains. I sometimes use the moisture of my breath to humidify small parts before gluing. The strength of the cyanoacrylate bond continues to increase slowly during the first 48 hours.

Shelf life of most cyanoacrylates is about 6 to 12 months—the glue thickens as it gets old—but storing the adhesive in the refrigerator will prolong its useful life. Since moisture speeds the setting time, however, allow the container to warm up to room temperature before you open it, or condensation in the bottle will offset the advantage of refrigeration.

Early cyanoacrylates did not work well on wood or other porous materials because the glue's viscosity is extremely low: the glue soaked into the surfaces, producing a starved joint. This same property, though, allows the adhesive to penetrate hairline cracks. Cyanoacrylate dripped into a ragged break will reinforce the fracture, and it can strengthen joints already glued with another adhesive that have loosened slightly if it is dripped along the existing glueline. You can repair a large gap by packing it with baking soda, then dripping glue on, which turns the powder into a hard, white filler. To fill a small crack or hole, put a few drops of glue in the crack, then sand immediately with wet-or-dry sandpaper before the glue sets. The wood dust mixes with the glue to form a hard filler that matches the color of the wood. This works best on dark woods; on light-colored woods the patch will be slightly darker than the surrounding wood. On thin stock, it's a good idea to put a piece of masking tape on the back of the crack, to keep the glue from sticking to the bench.

Cyanoacrylates do not set as quickly if there is acid present in the joint. If you're gluing an acidic wood such as oak, you'll get better results if you prepare the surfaces by brushing on a surface activator. These activators are mildly basic and neutralize the acid.

Cyanoacrylate is the most expensive adhesive generally available. Like epoxy, most brands are purchased in bulk and repackaged for retail sale, but you can save by buying in larger quantities. C.F. Martin Co. (of guitar fame), 510 Sycamore St., Nazareth, Pa. 18064, sells 3M Scotchweld CA-3 in 1-oz. containers for $12, Scotchweld Surface Activator for $8, or a kit consisting of two ounces of CA-3 and one container of surface activator for $25 (1984). Glue quality differs little between brands, but you will find a difference between the containers the glue is packaged in. Some styles clog before the adhesive is gone, when glue solidifies near the tip. Polyethylene dropper bottles are less likely to clog than metal squeeze tubes or rigid plastic containers. It helps to clear the nozzle by squeezing a little air out of the upright container before closing it.

Hot-melt glues

Hot-melt glues—Synthetic hot-melt glues are easy to apply and they set up quickly. Most are made of polyamide resins which melt at around 400°F. Hot-melts are widely used in industry, where their quick set is an advantage on assembly lines. Their good gap-filling properties make them ideal for repairing worn, sloppy joints in old furniture. Hot-melts form thicker gluelines than most other adhesives, and have relatively low strength and poor penetrating ability. They're good for temporary jigs or tack-on fastenings, where extreme strength is not required. They're also well suited for joints that may need to be disassembled, but the heat necessary to break the glue bond may also damage the surrounding finish. Hot-melts develop 90% of their final bond strength within 60 sec-

onds. The glue remains somewhat flexible and does not sand well. When the glue has cooled, excess can be removed with a sharp blade.

Hot-melt sticks are sold for use in an electric glue gun. Manufacturers make several grades that cool at different rates. Those sold in retail stores allow you only about 10 seconds to assemble parts, but you can increase open time slightly by preheating the parts. Hot-melts are also sold in thin sheets for veneering (available from Woodcraft Supply). You can use an ordinary household iron to provide heat, then weight or clamp the veneer until the glue has cooled.

Contact cements—Contact cements are rubber-based (usually neoprene) liquids that dry by solvent evaporation. They are used most often to bond high-pressure plastic laminates, such as Formica, to plywood or particleboard, without the need for clamps or prolonged pressure. Contact cements are sometimes used to attach veneers, but the glue bond can fail in spots because of seasonal moisture changes in the veneer, causing bumps in the veneered surface or separation at the edges.

There are two types of contact cements: solvent-based and water-based. Solvent-based cements, most of which are extremely flammable, dry in about 5 to 10 minutes. The nonflammable solvent-based cements are made with chlorinated hydrocarbons, and their vapors are toxic. These vapors are not trapped by an organic-vapor respirator, so you should use adequate ventilation with this and any solvent-based adhesive. Water-based contact cements are nontoxic and nonflammable, but they take longer to dry—about 20 minutes to an hour before parts can be assembled. The uncured adhesive is water-soluble, so you can clean your tools in water if the glue hasn't dried. Water-based contact cement provides better coverage than the solvent-based type, but it should not be used on metallic surfaces.

Contact cements are heat-resistant and water-resistant, although adhesive strength is low and the pliable glue film is likely to creep under load. Both types can be applied by brush, roller or spray. Apply adhesive to both surfaces to be mated and assemble when dry, but be sure that the parts are properly aligned. Adjustment is impossible once the two surfaces contact. Go over the glued surface with a roller to ensure an even bond.

"Construction adhesive" is a thick mastic used by carpenters to fasten flooring or wall paneling. As yet, it has not been widely used in other areas of woodworking.

A few other adhesives that may have limited use in woodworking are the acrylic cements, most commonly encountered as pressure-sensitive contact adhesives. Liquid acrylics are also used in some linoleum cements and other mastics. "Anaerobic adhesives" remain liquid when exposed to air and solidify when deprived of oxygen. The Loctite Corporation markets a variety of anaerobic adhesives which are widely used for securing nuts, bolts and threaded studs. Because of its high porosity, however, wood contains too much oxygen to allow anaerobic adhesives to set. □

Dovetail Jigs
We test three fixtures for routing carcase and drawer joints

by Paul Bertorelli

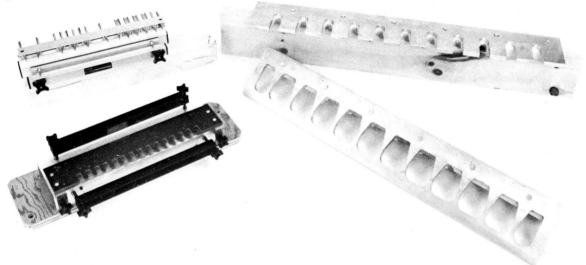

Router dovetail jigs are available in three types. The Sears jig, left foreground, will cut fixed-spaced half-blind dovetails for drawers or small carcases. Behind it is the Leigh tool, which will cut variably-spaced through dovetails. Keller's two-piece aluminum template, right, is designed for routing large dovetails in carcases. Pin spacing can be varied by shifting this tool.

Machine-cut dovetails have always gotten a bum rap. They're either too fat or too thin, or the angle is off, or something else is wrong that only cutting them by hand seems able to avoid—provided you've got the time and skill. These complaints, valid or not, have inspired the invention of router jigs that attempt to solve the problem.

Three basic types of dovetail jigs are now on the market—a large one for cutting through dovetails in carcases, and two smaller ones for through and half-blind dovetailing of drawers and smaller carcases. We bought one of each and I spent several days trying them out.

I wasn't surprised that I had to fuss to get the jigs to work well, but once router and jig have been accurately set up using test pieces, all three jigs will cut joint after joint with good results, providing that the stock has been planed accurately and cut off squarely. I was surprised, however, to find that I could join boards as well with the cheapest jig as with the most expensive.

Two of the jigs tested—David Keller's $325 two-piece aluminum tool and Leigh Industries' $149.50 adjustable device—are recent inventions. They

were designed partly to address the common complaint that machine-cut dovetails lack visual excitement because the angle and spacing of the pins and tails don't vary. With either of these two jigs, the pin angle is determined by the router bit, but you can change the look of the joint by varying the pin-tail proportions and spacing. Both of these jigs cut only through dovetails. Keller's is for large carcase joints, while the Leigh jig is for drawers. Sears' jig, the third type I tested, cuts only half-blind dovetails, and the space between pins and tails is fixed. At $45, its design and price are representative of jigs by at least three other manufacturers—Porter Cable ($68), Black and Decker ($68), and Bosch ($82), all of whom sell through local distributors. Bosch also makes a large jig (for $94) that will dovetail boards up to 16 in. wide (1982 prices).

Keller jig—David Keller, a Bolinas, Calif., woodworker, began selling his dovetail jig in 1976. When first conceived it was made of Plexiglas, and later of phenolic plastic. The version he sells now is made of two ½-in. thick, 36-in. long, machined aluminum plates—one

for the pins and one for the tails. A pair of bearing-equipped ½-in.-shank router bits (a 1-in. diameter, 14° dovetail bit and a ¾-in. straight bit) are included in the jig's $325 price (1982). The templates can be repositioned, so there's no limit to how wide the stock can be. Boards ranging in thickness from ⅝ in. to about 1¼ in. can be joined. The dovetail bit, however, has a limited depth of cut, so for stock thicker than ⅞ in., the pin board must be rabbeted.

To use this jig, you need a beefy router with a collet that will mount ½-in.-shank bits—I tried a 2-HP Milwaukee and a 2¾-HP Makita. The templates are first screwed to backing boards. These give you a way to clamp the tool to the work, and they keep chunks from being torn out of the stock as the bits exit from the cut. If the jig's built-in 3-in. spacing is used, it's simple to lay out the joint. The tail-cutting template is clamped to the end of the board, and the tails are milled with the dovetail bit. The template is removed and the tail locations are scribed directly to the pin board, as in hand-dovetailing. It isn't necessary to scribe each tail. If two or three are located accurately, the jig automatically lo-

A near-perfect joint is quickly attainable with all three of the jigs tested. Above foreground are the large dovetails with pins on 3-in. centers made with the Keller jig. The joints made with the Sears fixture, left, are of fixed spacing, giving them the unmistakable look of machine-made dovetails. With its variable pin spacing, the Leigh jig makes dovetails, above right, that come closest to looking handcut.

The Keller jig is furnished with bearing-guided router bits that follow the template more accurately than the guide bushings used with the other two jigs. This photo shows how the bearing guides on the template to cut the pins. And it also illustrates how close the bit comes to the templates, a condition that requires constant vigilance when using any of these jigs.

cates the others. The straight bit is used to cut the pins. If you prefer, the fixed spacing can be ignored and you can put the pins anywhere you like. In that case, the templates must be moved after each cut, and all the tail locations must be scribed to the pin piece, a tedious and inaccurate routine. You might as well dovetail by hand.

To tighten the joint, you make wider pins by moving the edge of the template toward the work; moving it away shaves them down and loosens the joint. Keller suggests you experiment with mounting the template on its backing board until a good fit is produced. But I found it handier to set the jig up to make a tight joint, and then loosen it with masking-tape shims.

Of the three jigs tested, Keller's was by far the simplest to use. Once I had it set up correctly, I could make tight dovetails that were attractive, but somewhat square and clunky-looking for my tastes. I like to start and end a dovetail

series with a half-pin, and by departing from the jig's fixed spacing I was able to do that, with only a minor loss of accuracy. The alternative is to let the pins fall where they may, as in the photo above, or to design in widths that are multiples of the jig's 3-in. spacing.

Avoiding tearout with this jig requires some care. After the backing board has been used a few times, too much material is cut away for the board to offer much protection as the bit exits. Slower cutting, or a new board, helps.

This jig does its job, but its price and size limit its appeal. Keller says it is aimed at small production shops that don't have the time to hand-dovetail carcases, or the capital to buy specialized machinery. The weekend woodworker should be wary—dovetailing with this jig calls for experience with large routers. I was reminded rather violently of the risks. As I was finishing cutting the pins on one test piece, I inadvertently tilted the router. The bit grabbed the

corner of the jig. The confrontation snapped the router's shaft and hurled the collet and bit out the bottom of the machine, shattering the Bakelite base as it went. I suffered a shrapnel wound in the face. While I don't see this jig as being more dangerous than other power tools, it does demand undivided attention and careful movement, along with good eye and face protection.

Keller's jig is available direct from him at Star Route, Box 800, Terrace Ave., Bolinas, Calif. 94924.

Leigh jig—Ken Grisley, an English boatmaker now living in Canada, designed a jig that will make pins and tails with variable spacing. Earlier this year, he formed Leigh Industries Ltd. to manufacture and market this device, the only small jig we could find that can make randomly spaced, through dovetails.

The Leigh jig consists of ten pairs of movable, die-cast aluminum fingers mounted on a heavy aluminum extrusion. The boards to be joined are clamped to the jig, pin board on one side, tail board on the other. The aluminum fingers project over the stock to guide the router bit through the cut. The fingers can be positioned anywhere along the extrusion and are locked in place with socket-head screws. The pin side of each finger pair is angled at 15° to match the dovetail bit used with the tool (bits aren't included in the price). The opposite or tail side of the fingers is straight. The jig is intended for making drawers, since it works best cutting through dovetails in stock up to ½ in. thick and 12 in. wide. But by cutting a rabbet in the pin board, it's possible to dovetail stock up to ¾ in. thick.

Instead of bearings to guide the bits through the wood, a guide bushing is attached to the router base and the bit is centered inside the bushing's bore. Bushings tend to be less accurate than bearings because it's difficult to get and keep the bit concentric.

Setting up the jig involves loosening and retightening 20 socket-head screws (two for each finger pair) to achieve the desired spacing. It takes time. For the effort, though, you get pins and tails where you want them, and you can start and end the joint with a half-pin, no matter what the width. You can also vary the width of pins and tails. I used a 1-HP Sears router and ¼-in.-shank carbide bits to test the Leigh jig: tails first with the dovetail bit, then pins with a

$\frac{5}{16}$-in. straight bit. Since the fingers automatically locate the pins on the piece clamped in the other side of the jig, no scribing is necessary. Leigh supplies a stack of $\frac{1}{64}$-in. paper shims for adjusting the tightness of the joint. The shims fit under an aluminum plate to which the pin board is clamped. Removing shims moves the work closer to the jig, thus tightening the joint.

After some confusion over finger spacing, I was able to cut tight dovetails in both hardwoods and softwoods. Once I got the hang of it, changing the spacing was easy. The biggest problem I had with this jig was tearout. As the bits leave the cut, a ragged edge often results. Backing the cut with a strip of wood clamped between the work and the jig corrected this, but I had to replace the strip frequently.

For all its complexity—it has some 25 separate parts—the Leigh device held its adjustments throughout the day I used it to join a batch of drawers. The last corners I made were as tight as the first. The tool seems robust enough for a small production shop. But it may be pricey for the woodworker with just a few drawers to dovetail; and besides, in the time it takes to adjust the jig, you really could chop quite a few hand dovetails. Moreover, I found that having to rabbet the pins in stock thicker than $\frac{1}{2}$ in. limits the tool's versatility.

The jig is sold through mail-order tool distributors; Leigh Industries (Box 4646, Quesnel, B.C., Canada V2J 3J8) has a full list.

Pin spacing can be varied with the Leigh jig by moving fingers locked down by socket-head screws. Fit of the joint is determined by the pins' width, which is controlled by adding $\frac{1}{64}$-in. thick paper shims stacked behind an aluminum plate between work and jig. The plate at left fastens over fingers to give the router a smooth surface to ride on.

The Sears dovetailer is the quickest to use because both pins and tails are milled in one operation with the same bit. The workpieces are clamped together, limiting tearout and making joints consistently clean.

Sears jig—Singer Motor Products Division has made the Sears dovetail jig for more than 20 years. It consists of a phenolic template machined with fingers and mounted on aluminum channel iron. The pieces to be joined are clamped at right angles to each other under the template. The router (through a guide bushing) cuts half-blind pins and tails in a single operation with one cutter—a $\frac{1}{2}$-in. diameter dovetail bit. Boards up to 12 in. wide and from $\frac{3}{8}$ in. to about 1 in. thick can be joined. The dovetails made by this jig are square-dimensioned and closely spaced, with tails undercut on the inside face. Another template sold separately can be fitted to the jig to make $\frac{1}{4}$-in. half-blind dovetails in stock as thin as $\frac{5}{16}$ in. Though intended only for half-blind dovetails, the Sears fixture can also cut through dovetails. It requires routing perilously close to, or

even into, the aluminum base of the tool, however. And the ill-fitting, half-round pins thus produced are hardly worth the effort.

Pin and tail spacing is fixed with this jig, so set-up involves little more than ripping the stock to width and squaring the ends to be joined. The critical and most difficult task is getting the joint tightness just right. You do it by changing the router's depth of cut, lowering the bit to tighten the joint, raising it to loosen it. With the crude rack-and-pinion depth control on a Sears router, adjustments take lots of trial and error. A router with a better depth control, I suspect, would simplify this job.

Once set correctly, though, the Sears jig is faster and easier to use than the other two because the 15° dovetail bit mills pins and tails at the same time. No

bit or jig changes are needed. The boards are clamped tightly against each other during the cut, thus backing the bit's exit and eliminating tearout.

The performance of this jig surpassed my expectations. I once bought a cheaper version, and after using it twice with mediocre results, tossed it back behind the scrap pile. The version I tested, though, is better made yet still priced low—well within the budget of most woodworkers. True, the pin-tail spacing can't be varied, but the precise regularity of the joint made by this jig has a certain appeal. It looks exactly like what it is—a machine-made dovetail. □

Cutting Dovetails With the Tablesaw
A versatile way to join a stack of drawers

by Mark Duginske

For joining such basic casework as small boxes, chests and drawers, I've always felt that there was a missing link between the tedium of hand-cutting dozens of dovetails and the faster method of producing monotonous-looking joints with a router jig. With that in mind, I developed this table-saw dovetail method which combines hand-tool flexibility with power-tool speed and accuracy.

With this technique, you can vary both the width and the spacing of the pins and tails for practically any aesthetic effect. The blocks that set the spacing are self-centering and will produce perfect-fitting, interchangeable joints, eliminating the need to mark boards so that individual joints will fit, as with hand-dovetailing. Besides a good combination sawblade and dado head for your tablesaw, you'll need a marking gauge, a bevel gauge and a couple of sharp bench chisels. Before proceeding, screw a wooden fence to the saw's miter gauge. A 3-in. by 20-in. fence will safely support most work.

Begin by squaring the ends of the boards to be joined. Take your time with this step—inaccurately prepared stock virtually guarantees sloppy results. I spaced the pins equally for the 4¼-in. wide drawer parts I'm joining in the photos. You can mark the pin centers directly on the pin boards, or, as I did here, you can just cut the spacer blocks to create whatever spacing you want the pins to have. In any case, the width of the blocks should equal the distance between pin centers. You'll need one block for each full pin, plus one.

The pin size is also controlled by the blocks. When they're lined up edge-to-edge, the total width of all the blocks should be less than the width of the stock by an amount equal to the width of the narrow part of each pin, that is, on the outside face of the pin board. I chose ¼-in. pins for the drawer sides shown in figure 1; if you want finer pins, decrease this dimension. The blocks must be of consistent width, so I crosscut them from the same ripping, then sandpaper off any fuzzy corners so that they'll line up with no gaps. To mark the depth of the pin and tail cuts, set your marking gauge to the stock thickness, and scribe a line on the faces of the pin board and on the face and edges of the tail board.

Cut the tails first with the saw arbor (or table) tilted to 80°, an angle that I've found produces the best combination of appearance and strength. A bevel gauge set at 80° can be used to set both the sawblade for the tails and, later, the miter gauge for the pins. As shown in figure 2, position and clamp a stop block to the miter-gauge fence so that when all the blocks are in place, a half-pin space of the correct size will be cut. At its narrowest width, the half-pin space should equal the narrow width of a pin. Raise the sawblade until it cuts right to the gauge line, then, with all the blocks in place, begin cutting the tails, flipping the board edge-for-edge and end-for-end (photo, right). Continue this process, removing a spacer block each time, until all the tails are cut.

A good-quality carbide-tipped blade will saw crisp pins

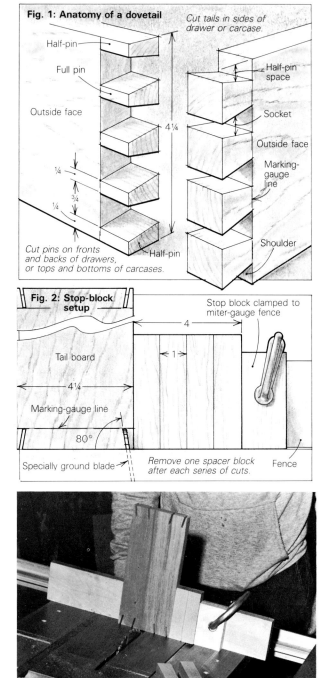

Fig. 1: Anatomy of a dovetail

Cut tails in sides of drawer or carcase.

Half-pin

Full pin

Outside face

Half-pin space

Socket

Outside face

Marking-gauge line

4¼

¼

¾

¼

Cut pins on fronts and backs of drawers, or tops and bottoms of carcases.

Half-pin

Shoulder

Fig. 2: Stop-block setup

Stop block clamped to miter-gauge fence

4

Tail board

1

4¼

Marking-gauge line

80°

Specially ground blade

Remove one spacer block after each series of cuts.

Fence

In Duginske's tablesaw dovetail method, the tails are made first in a series of cuts with the table or arbor set at 80°. After each series, a spacer block is removed and the cuts are repeated for the next tail. The last tail is made with one block in place.

Photos: Bill Stankus; drawings: David Dann

Machine-cut dovetails don't have to have the stiff, predictable look dictated by many router jigs. Using your imagination and the author's tablesaw technique, you can vary the width and spacing of pins and tails for infinite visual variety.

and tails, but set at an angle it leaves a small triangle of waste at the bottom of the cut that must be chiseled out later. To minimize handwork, I had the tops of the teeth on a carbide blade ground to 80°. The grinding cost $12 (1983) and the blade can still be used for other work. If you have a blade ground, make sure that all the teeth point in the same direction, and when you tilt your saw, match the tooth angle.

To cut the pins, clamp the boards together and scribe either of the outermost tails onto the pin board with a knife, as in the photo at right. Mark the wood to be wasted with an X. Only one pin need be marked; the spacer blocks will automatically take care of the others. The pins will be formed in the series of three cuts illustrated in figure 3.

First, return the arbor or table to 90° and install a ¼-in. dado blade raised to cut right to the gauge line. Adjust the miter gauge to 80°, and with all the spacer blocks in place, reset the stop block so that, with the outside face of the board positioned away from you, the first dado cut will be made just to the inside of the knife line. Make sure that the board is positioned correctly, or else you'll end up cutting the pin angle in the wrong direction. Make the first cut, flip the board end-for-end and cut only the opposite corner. Then remove the first block and repeat until one side of each pin is cut.

For the second series of cuts, set the miter gauge to 90° and waste the material between the pins. You'll have to remove a lot of wood in several passes to form widely spaced pins, in which case it's handier to judge the cuts by eye rather than relying on the spacer blocks. Don't waste too much material, else you'll nip off the opposite side of the pins. While the miter gauge is at 90°, use the dado blade to waste the wedge of wood remaining in the sockets of the tail board. If the sockets are narrower than ¼ in., nibble out the wedges on the bandsaw or with a coping saw and a chisel. Use a backsaw or the tablesaw to trim the shoulders where the half-pins fit.

Next set the miter gauge to 80° in the opposite direction, and reset the stop block so that the dado blade cuts just inside the other knife line. Make the third series of cuts like the first, but before proceeding, slip three or four strips of paper between the last spacer block and the stop block. Complete the cuts and try the joint. It should slip together by hand or with light mallet taps. If the joint is too tight, remove one or more paper shims, repeat the cuts and try again. Smooth the space between the pins and the tails, pare any tight spots with a chisel, and you're ready to glue up. □

Mark Duginske lives in Wausau, Wis.

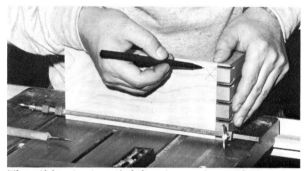

The tail location is scribed directly onto the outside face of the pin board with a knife. A bold X marks the material that will be wasted to form the pins.

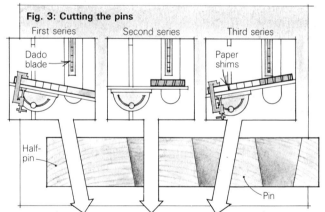

Fig. 3: Cutting the pins

First series   Second series   Third series

Dado blade   Paper shims

Half-pin   Pin

Once marked out, the pins are formed by wasting the wood between with dado-blade cuts. In the photo below, Duginske completes the second series of pin cuts.

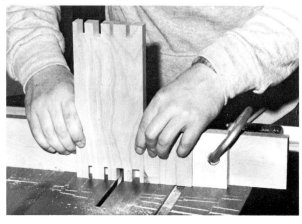

Curved Dovetails
Secret miter is the key

by John F. Anderson

Through dovetails with curved baselines can enhance case joinery and provide visual interest. Making pins and tails a different thickness from the stock being used involves cutting a curved secret miter on the inside corner of the joint.

The appearance of through dovetails is ordinarily determined by the thickness of the pieces being joined, and the exposed end-grain surfaces of both pins and tails usually make a straight line parallel to the corner of the joint. But this baseline need not be straight. By varying the thickness of the pins and tails, you can create curves along both sides of the joint. Curved dovetails open up new design possibilities. Incorporated into a cabinet, drawer or box, this joint can enhance and accent the overall composition.

The procedure for making curved dovetails is basically the same as for cutting regular ones. But there are two notable exceptions—you need a curved template block for laying out the curved baselines and for chiseling, and you have to cut a curved-bottom blind miter along the inner edges of the joint. You'll need a sharp pencil, a marking gauge, a dovetail saw, chisels and mallet, and a good eye.

The inside miter is the key to curved dovetails because it lets you make pins and tails that are thinner in section than the thickness of the stock and thus permits a wide range of curves to be used. They can be either concave, convex or reversed, symmetrical or asymmetrical, and the thicknesses of the pieces joined need not be the same. The curves on each side (pin and tail) need not be the same either, though to accomplish such a joint you need to make two templates and to lay out the curved cuts alternately (figure 5). The procedures described below will work for any curve, so long as the miter is cut properly.

First make the curved template block from a piece of wood at least ¾ in. thick. It should be large enough to be clamped firmly to the workpiece (without the clamp getting in the way) and precisely the width of your case sides, for ease of alignment. After laying out the curve, take care in cutting it, making sure that the curved end-grain edge is perpendicular to the bottom face. You'll use this edge to guide your chisel when making vertical cuts, so it has to be a true 90°.

Next, set a marking gauge to a little more (maximum 1/32 in.) than the thickness of the stock and score the inside faces of the work. The inside faces should be finish-planed or sanded to avoid knocking off the edges of the curves after the joint is completed. Don't strike a baseline across the outside face of the stock; rather, score marks at the outer edges that will allow you to position the curved template. Trace the curve from the template onto both faces of the stock, using the straight baselines for registration; also trace the curve onto the end-grain edges (figure 1). After careful consideration, mark out the pins first; I do this freehand, but you can use any means you like. Clamp the piece in a vise and saw down the waste side of each pin to the curved line.

Sawing done, lay the piece, outside face up, on the bench and clamp the template on top in its original position. Using the edge of the template as a guide for the back of your chisel (figure 2), chop into the waste deep enough so that no chip-

ping or tear-out will occur when you finish chiseling through from the other side. Turn the piece over, clamp the template into position on the inside face, and chop straight down, removing chips near the edge of the template with oblique paring cuts (figure 3). Don't go any deeper than the curved line marked on the end-grain edge, and don't try to remove the waste between the pins just yet. Clean up the bottom of the cut by paring in from the edge with a narrow chisel or in-cannel gouge. This must be done acurately because the results will determine the fit of the miter on the inside corner. Now remove the template, secure the workpiece again to the bench and cut this miter using a sharp chisel, beginning at the baseline and paring to the intersection of the vertical shoulder and the curved bottom (figure 4). This will automatically produce a miter at the proper angle.

Now that the miter has been cut, return the template to its former position and use it as a guide for removing the waste between the pins. Cutting the miter comes before this because the entire bottom curve is still visible then, making it easier to cut to this line, which is partially obliterated when the waste between the pins is removed.

Now lay out the tails from the pins as in standard practice. Saw down on the waste sides of the tail lines to the curved line. Then simply repeat the procedures used for mitering the inside corner described above, chiseling out the waste between the tails last. The pieces should fit together correctly, leaving no gap on the inside of the corner.

John Anderson lives in Bottineau, N. Dak.

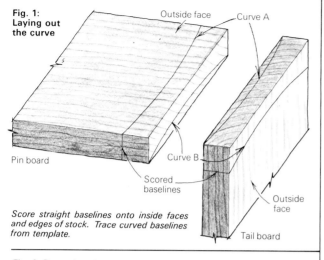

Fig. 1: Laying out the curve

Outside face

Curve A

Pin board

Curve B

Scored baselines

Outside face

Tail board

Score straight baselines onto inside faces and edges of stock. Trace curved baselines from template.

Fig. 2: Preparing the outside face

Template clamped on stock

Outside face

Baseline

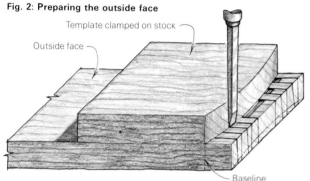

Using template edge to guide chisel, sever the waste tissue across the grain to prevent tear-out when chiseling from the reverse side. Note that tail board (photo below), which will be marked from completed pin board, is prepared similarly.

Fig. 3: Cutting the inside curves

Inside face

Paring cuts with grain

Vertical shoulder

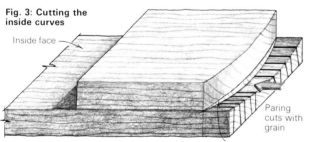

Again using template as a guide, cut the curved vertical shoulder and pare away waste along curved line on inside face of pins. Photo below shows tail-board shoulder similarly cut.

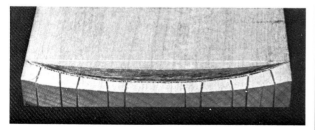

Fig. 4: Cutting the inside miter

Scored baseline

Vertical shoulder

Curve A

Curve B

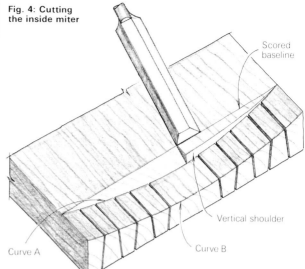

Cut the miter from straight baseline to curved one. Remove waste between pins only after cutting miter.

Curve A and curve B need not be the same. If they are, as in the example above, then the miter is a true 45°. If curves A and B are dissimilar, as in the drawing below, then the angle of the miter will vary along the length of the joint, like a slightly twisted ribbon.

Fig. 5: Laying out two different curves

Curve A

Pin board

Curve B

Tail board

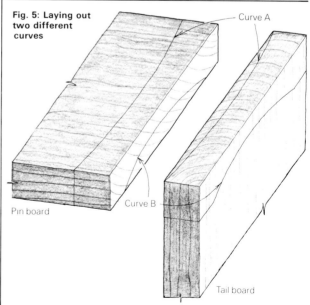

The Butterfly Joint
Double dovetails for strength and beauty

by Frank Klausz

Though the butterfly joint, sometimes called a double dovetail, is ancient, it was hardly ever used in traditional furniture making. Lately, however, it has enjoyed wider use since people have begun to make furniture from solid slabs of wood, from whole flitches or from root sections. Some of George Nakashima's tabletops, for example, show how the butterfly can be used for strength and decoration. Across the grain, this joint provides mechanical reinforcement and is especially useful for controlling checks in slabs and for repairing cracks in tabletops and chair seats. The joint can also be used to join separate boards into a single panel or to join up sections that are butted together lengthwise to form a long tabletop or bar top. Recently I put butterflies into a horseshoe-shaped kitchen countertop made from six separate pieces of butcher-block material. These were joined end-to-end with butterflies, three of them for each joint.

I try to make all my butterfly keys the same size, unless the job demands otherwise. So most of the time I cut them 3¾ in. long, 1½ in. wide and ⅝ in. thick. Instead of cutting them one at a time, I like to make a dozen at once. I first cut several strips of wood to a width of 1¾ in. and a thickness of ⅝ in. Then using my radial arm saw and a stop gauge, I cut the strips into pieces exactly 3¾ in. long. I glue 12 of these pieces together, face to face, with just a spot of glue in their centers. I wrap masking tape around the end grain, then I clamp them firmly together. I tilt the arbor on my table saw to a 10° angle and set the fence precisely 1½ in. from the blade where it intersects the plane of the table. The blade should be set 1¹³⁄₁₆ in. high (or slightly less) so the two waste pieces will stay attached to the stock after all four cuts. Make sure you leave an unsawn strip at least ³⁄₁₆ in. wide in the cen-

ter of the wood. With the clamp positioned so it can't contact the blade, I make the cut with one hand, using the bar of the clamp as a handle while pressing down on the stock. It is a hair-raising operation for one who hasn't had much experience using a table saw, but with care it can be done safely.

After making the four necessary cuts, you end up with a stack of perfectly dimensioned butterflies, except for the two waste pieces still attached to the center on both sides. Break these pieces out and clean up the valley by passing a sharp chisel left and right. Now knock the individual butterflies apart. The whole job from gluing to knocking apart shouldn't take longer than a half-hour, or the glue will set hard and you'll have a solid block of wood. Then you'll have to saw the pieces apart.

Instead of gluing 12 or more pieces together, you can cut them all from a single, solid block, if the grain runs in the right direction. This method is economical because you can cut off pieces of whatever thickness you need and save the rest. It also allows you to use short trimmings from wide, thick planks, pieces that ordinarily would be thrown away. Cutting all the keys from one piece is safest because you don't have to clamp the workpiece while sawing and you don't have to worry about them separating during the cut.

If you need only one or two keys, it's best to make a pattern, trace it onto your wood and bandsaw close to the line. Then clean up and straighten the edges with a sharp chisel.

Once you've made the butterflies, you have to set them into the wood. The simplest way is to place a key on the surface (centered across the check or joint and perpendicular to it) and trace around the butterfly with a sharp pencil. With a chisel cut the mortise, being careful to get no closer to the

Butterfly keys are an attractive way to control checks. They can also be used to join single boards into panels.

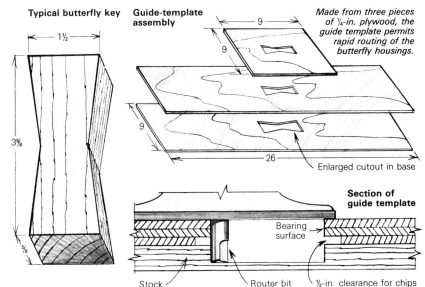

Typical butterfly key

1½

3⅝

⅝

Guide-template assembly

9

9

9

Made from three pieces of ¼-in. plywood, the guide template permits rapid routing of the butterfly housings.

26

Enlarged cutout in base

Section of guide template

Bearing surface

Stock

Router bit

½-in. clearance for chips

Photos: Frank Klausz and Edward Ludlow; drawings: Ric Lopez

Cutting a dozen keys at one time on his table saw, Klausz uses a bar clamp to help grip the stock. This is a dangerous cut. Do it safely by making sure that the blade clears the clamp jaws and that the thumb is placed high on the workpiece.

To clean up the unsawn center portion of the stock, Klausz holds the chisel askew and slices at the waste, moving from left to right.

lines than ¹⁄₁₆ in. The floor of the cavity can be leveled and smoothed with a router plane and then the walls pared to the line with a chisel.

Because I make a lot of them, I prefer to cut these mortises with my electric router, using a guide template that makes the job quick and easy. So I can make the cutout in the template exactly the same size as the butterfly, I use a two-flute, ¼-in. router bit, which lets the shank bear directly on the template guide. The best bit for this has cutting edges only ½ in. long and is made by Velpec (#4-4-AI) available from Force Machinery Co., Rt. 22, Union, N.J. 07083. You can use other ¼-in. bits, as long as their cutting edges are flush with the shank, but since most of these bits (Rockwell and Sears) have cutting edges longer than ½ in., you'll have to increase the depth of your template so the shank will ride smoothly on the cutout edge with no danger of the bit cutting into the template itself. Remember, the farther the cutting edge is from the router chuck, the more chatter produced and the rougher the cut.

To make the guide template you will need three pieces of good-quality ¼-in. plywood, two of them 9 in. by 26 in. and one 9 in. square. Glue the small square piece in the center of one of the longer pieces. Then lay one of your butterflies on the center of the square piece and draw around it with a sharp pencil. Drill a pilot hole inside one of the corners and saw out the waste with a saber saw, or you can chisel it out. Make sure you are doing very good work here, for all your butterflies will fit like this one. Leave your pencil line on the wood so you can do the final fitting with a file for a perfectly snug fit.

Now take your third piece of plywood, put it beneath the guide template just made and align the outer edges. Trace the cutout onto the bottom piece, remove the guide template and enlarge the tracing by ½ in. all around. Cut out this area with a saber saw. This bottom piece elevates the template the right amount; the enlarged cutout makes room for chips and dust that would otherwise interfere with the bit as its shank bears against the sides of the template. It's a good idea to glue a couple of sandpaper strips to the bottom of the base to help keep it from slipping. Put a little oil on the pattern where the router bit or guide bushing will rub.

Remember that you want the mortise to be about ⅛ in.

shallower than the thickness of the butterfly so you can plane the key flush to the surface. For a butterfly that's ⅝ in. thick, I cut the mortise ½ in. deep.

If you want to use an ordinary straight-face router bit and a guide bushing, then you'll have to make the cutout in your template larger than the butterfly. The exact amount of this enlargement depends on the distance from the cutting arc of your bit to the outer edge of your guide bushing (collar). Try using a setup that will cut a template ⅜ to ¾ in. larger than the pattern, as this is usually manageable for inlays; you will be treating the key like an inlay, dropped in and planed flush.

To rout out the waste, hold the router at an angle so part of the base contacts the surface of the template and center the bit over the cutout. Switch on the power and let the router down flat. Work the tool in a clockwise direction, going from the center to the outside edges, and when you've removed the waste from the center, make a final pass around the edges so that the shank of the bit (or guide bushing) rubs the edge of the template. All that's left to do is to chisel out the four corners, where they've been left rounded by the router bit.

Because the butterfly is usually housed across the grain and the greatest amount of shrinkage and expansion occurs in this direction, a slight undercut on the ends of the mortise is sometimes desirable. This will help prevent the wood from checking if it shrinks against the ends of the key. Butterflies used for repair purposes and visible only from underneath should be cut slightly shorter than their housings, leaving a small gap at either end in case the wood moves.

The butterfly should be glued into its housing, clamped if possible and allowed to dry before you plane it flush with the surface. Avoid using a belt sander to work it down because you can never get the cross-grain scratches out. I use a sharp smoothing plane, followed by a cabinet scraper. Then I sand with 220-grit paper.

Because this joint needs very little material, I try to use dark woods—ebony, rosewood, padauk, purpleheart and black walnut—for contrast with the lighter wood of the tabletop or counter. If, on the other hand, the table is made from a dark wood, I use a light wood for the butterflies, such as lemonwood, satinwood, curly maple or white ash. □

Bandsawn Dovetails

Tilt, saw and chop

by Tage Frid

Routers and tablesaws aren't the only way to make through dovetails with a machine. I use my bandsaw to cut the pins and tails, and the results aren't much different from cutting the joint by hand. Start by using a marking gauge to scribe the baselines of pins and tails on both boards. Cut the pins first. Tilt the bandsaw table 10° (or whatever angle you wish your pins to be) to the right, and clamp a fence parallel to the blade and slightly farther away from it than half the width of the stock, as in drawing **A.** (If your table won't tilt in both directions, build a clamp-on auxiliary table to reverse the work.) Clamp a stop to the fence so that the blade will cut just to the baseline. All the cuts for this method should be made with the inside face of the board up. Mark your stock so that you won't lose track. Make the first cut, which will be one side of a center pin, then turn the stock end-for-end and make the second cut, one side of the other center pin.

Between the stock and the fence, place a spacer equal in width to the pin spacing. For this example, you'll cut one center pin and two half-pins at each edge. Cut the half-pin on one end of the stock, then turn the board end-for-end and cut the other half-pin (**B**).

Now tilt the table 10° to the left, move the fence to the opposite side of the table, and use the spacer to cut the other two half-pins (**C**). Then remove the spacer and cut the other side of the two center pins (**D**). Chisel out the waste in the pin boards, just as you would in making hand dovetails.

With the pins chiseled out, scribe their location directly on the tail board (**E**). Return the bandsaw to the horizontal position and saw freehand to the waste side of the lines that mark the tails. To remove the waste where the center pins will fit, saw up to the baseline repeatedly (**F**), shifting the stock sideways each time, before cleaning to the line with a chisel. To waste the area where the half-pins will fit, saw right up the baseline (**G**). Try the joint and adjust its fit with a chisel where necessary. This method will work with wider boards, but you'll need more spacers to locate the other pins. □

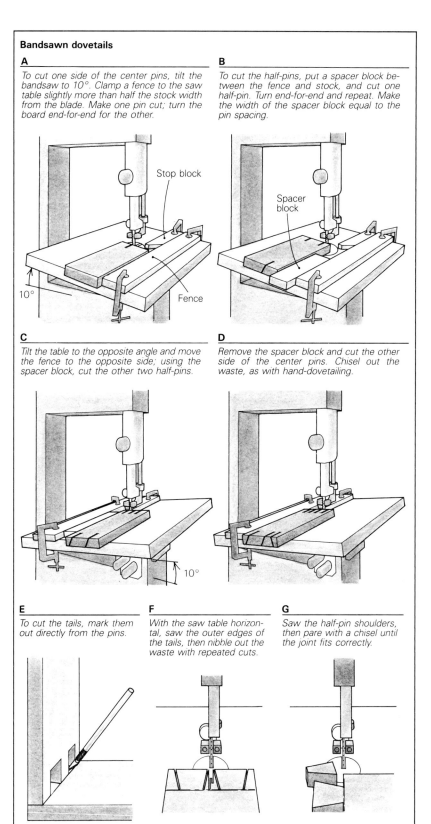

Bandsawn dovetails

A

To cut one side of the center pins, tilt the bandsaw to 10°. Clamp a fence to the saw table slightly more than half the stock width from the blade. Make one pin cut; turn the board end-for-end for the other.

B

To cut the half-pins, put a spacer block between the fence and stock, and cut one half-pin. Turn end-for-end and repeat. Make the width of the spacer block equal to the pin spacing.

C

Tilt the table to the opposite angle and move the fence to the opposite side; using the spacer block, cut the other two half-pins.

D

Remove the spacer block and cut the other side of the center pins. Chisel out the waste, as with hand-dovetailing.

E

To cut the tails, mark them out directly from the pins.

F

With the saw table horizontal, saw the outer edges of the tails, then nibble out the waste with repeated cuts.

G

Saw the half-pin shoulders, then pare with a chisel until the joint fits correctly.

Drawing: David Dann

A Two-Way Hinge
Careful routing makes screen fold

by Tim Mackaness

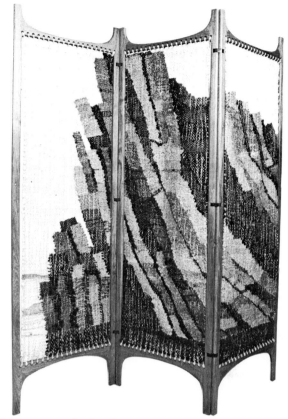

Many folding screens must be able to fold up like a concertina as well as stand alone as a triptych. Traditional hardware companies offer few acceptable hinges for the two-way folding screen. An attractive and functional concertina-type screen hinge can be made of wood by a sequence of careful but simple router cuts.

The hinge, visually symmetrical, can be de-emphasized if made of the same wood as the screen, or exaggerated to produce an interesting design detail by choosing a wood of contrasting color. Either way, a strong wood must be used. This design may be adapted to screen stiles of almost any thickness. The screen shown is made of 7/8-in. stock. I've found that wood thinner than 3/4 in. is too fragile. In any case, remember that the hinge pivots about the center of the dowel pins, which are concealed by inlays.

The use of round-over bits to produce a machine fit for the inlays and to radius the hinge member is handy and yields a very precise piece, but equally fine results can be achieved by carefully rasping and sanding the radius by eye. A practice hinge made before you confront the actual screen will build your confidence, let you determine the proper tolerances and provide a handy crutch for future hinges. □

Folding screen of teak and rosewood made by author, with 'Columnar Basalt' tapestry by Judy Nylin. Dowels reinforced with mitered glue blocks join rails and stiles. Curve is bandsawn after assembly, then template-routed to make all three panels identical.

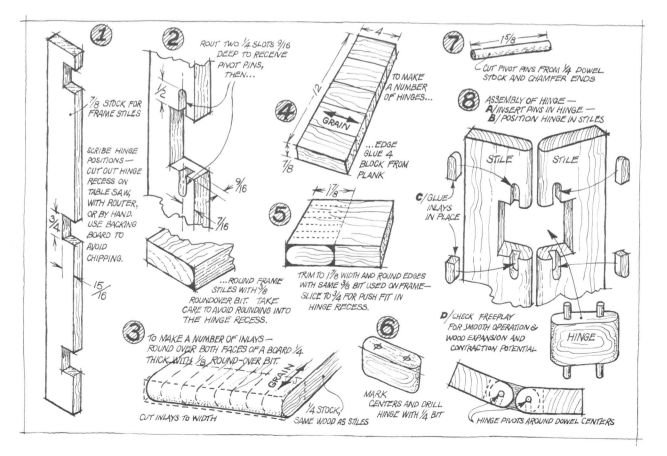

Knockdown Furniture
Form follows junction

by Curtis Erpelding

Although I design and make furniture for private clients, my pieces usually incorporate knockdown features. I like to devise designs suitable for mass production because I believe that functional, well-crafted furniture should be affordable. Because assembly time is eliminated, knockdown furniture is economical to produce commercially, and the pieces can be shipped disassembled, in compact packages. This economy and convenience are particularly relevant now, when people live in smaller spaces and move frequently. But the heart of my interest in knockdown design lies beyond function or practicality.

What interests me fundamentally is the concept that machined pieces of wood can lock together into a new form in ways that take into account the wood's physical properties. The basic problem becomes how to join (and later separate) two pieces of wood using gravity, friction and, occasionally, metal fasteners.

In 1980, the Design Arts Program of the National Endowment for the Arts (1100 Pennsylvania Ave. N.W., Washington, D.C. 20506) awarded me a one-year project grant to explore knockdown wooden furniture design. I had to satisfy three grant requirements: First, I was to continue my work in applying knockdown design to standard household and office furniture. Second, I was to develop three knockdown prototypes that would be suitable for mass production. And third, I was to research the possibility of having my designs commercially produced.

The three prototypes I decided on were a platform bed, a stacking chair with a circular seat, and a shelf system that leans against the wall. Each presented challenging design problems. In the beginning, I had intended to make detailed drawings and mockups to establish dimensions and proportions before building the actual prototypes. This goes against my usual practice. Normally I start with a fairly firm idea, work out the joinery details hastily on paper, cut a practice joint, and then plunge right into building the piece. When it works, this method saves a great deal of time. When it fails, the results are disastrous. Each procedure must be done correctly the first time, and design changes must be anticipated in advance, or they will interfere with the already completed part of the project. Sometimes I'm stuck with a half-completed piece of furniture and the prospect of having to start over again. For the grant project, I wanted to avoid this by doing drawings and mockups. Intentions may be noble, but bad habits die hard. As much as I tried, I kept reverting to my usual method. Somehow, for me, the design process has to involve this element of immediate risk. Either get it right the first time or blow the whole project.

Of the three projects, the platform bed, designed for a Japanese futon mattress, most closely approaches my notion of a pure knockdown design because it is built without glue or

There's no glue and no screws in Erpelding's knockdown platform bed: well-designed joinery keeps all the pieces in place. A router-cut dovetail joins each bed slat to the side rail. A dowel driven into the floor rail prevents the wedge from breaking the weak short-grain at the corner.

Photos: Joseph Felzman Studio, except where noted

metal fasteners. The rail-to-rail joint locks the parts and circumvents a troublesome characteristic of traditional bridle joints: an increase in humidity causes wood to expand across the grain, locking a tight-fitting bridle joint. This is fine for conventional furniture, but I had learned in earlier experiments with bridle-joint construction that knockdown structures would freeze solid in the middle of a humid summer. To solve this problem, I modified the bridle joint by tapering it (figure 1). This joint tightens under load, yet because of its geometry, it easily loosens when force is applied in the opposite direction. Simply wiggling one of the rails, or tapping upward lightly, breaks the assembly.

The tapered bridle joint wouldn't hold the pieces together by itself, so I added a wedge that fits into an angled notch in the floor rail. This wedge, which has a compound taper, locks the joint by forcing the side rail tightly against the side of the notch in the floor rail. The bed can be lifted and carried by the side rails, and the joint won't slip.

I cut most of the modified bridle joint on the radial-arm saw. To accommodate expansion, I cut both parts of the joint slightly wider than the thickness of the rails. The tapered face in the floor-rail part was surfaced with a router on an angled jig, and I feathered all sharp edges of the joint with a chisel.

The round cap that protrudes from the top edge of the floor rails is the end of a dowel, driven 4 in. into the rail. The dowel strengthens the weak short-grain, keeping the wedge from breaking it off.

In keeping with the solid-wood construction, I opted for slats, rather than plywood, to support the mattress. The soft, cotton futon mattress is used without a box spring, and traditionally it's unrolled right on the floor. The slats had to be thick enough and spaced closely to provide a firm platform for the futon, yet not so close as to interfere with ventilation. The obvious solution was to cut individual pockets in the side rails for the ends of each slat, while leaving enough space for the slats to expand without locking. I tapered both the sides of the pockets in the rails and the dovetails on the slats 15° so that the joint can swell without locking. I cut the joint with a 1-in. dovetail bit in a router using a jig that cuts both the pockets and the dovetails (figure 2). To cut the slats, I used a router with a smaller-diameter base.

The skirts at the head and foot of the bed are installed after the bed is assembled but before the wedges are driven in. The ends of each skirt are stop-slotted and slip over splines in the side rails. The outside face of the skirts is planed concave to match the radius on the ends of the rails.

With the bed, I suppose, I drifted furthest from the grant requirement that my design prototype be manufacturable. I don't pretend for a moment that this bed could be easily manufactured—although with simpler joinery it could, conceivably, be a production prototype.

While joinery was the major design feature of the knockdown bed, this is not true of the three-legged stacking chairs. Here, most of the joinery is straightforward and expedient—socket-head cap screws.

I was aware of precedents to this design. Finnish designer Alvar Aalto made spiral-stacking, circular-seat chairs, and Rudd International manufactures a similar stacking chair, but both of these designs have four legs and neither one knocks down. Hans Wegner designed a three-legged stacking chair, but it does not stack on the rotational principle. I felt that my

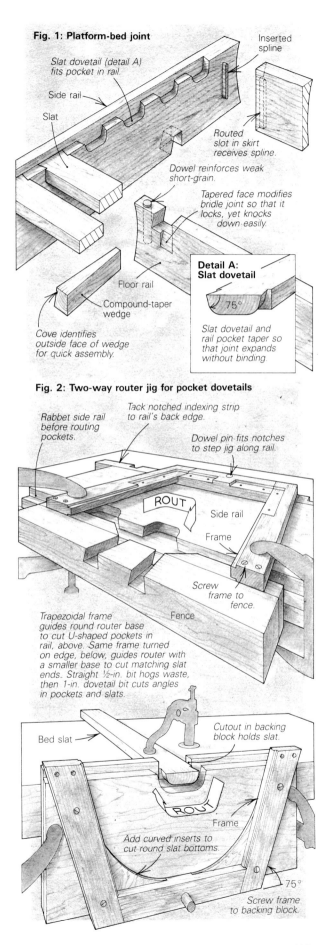

Fig. 1: Platform-bed joint

Slat dovetail (detail A) fits pocket in rail.

Side rail

Slat

Inserted spline

Routed slot in skirt receives spline.

Dowel reinforces weak short-grain.

Tapered face modifies bridle joint so that it locks, yet knocks down easily.

Floor rail

Compound-taper wedge

Cove identifies outside face of wedge for quick assembly.

Detail A: Slat dovetail

75°

Slat dovetail and rail pocket taper so that joint expands without binding.

Fig. 2: Two-way router jig for pocket dovetails

Rabbet side rail before routing pockets.

Tack notched indexing strip to rail's back edge.

Dowel pin fits notches to step jig along rail.

ROUT

Side rail

Frame

Screw frame to fence.

Trapezoidal frame guides round router base to cut U-shaped pockets in rail, above. Same frame turned on edge, below, guides router with a smaller base to cut matching slat ends. Straight ½-in. bit hogs waste, then 1-in. dovetail bit cuts angles in pockets and slats.

Fence

Bed slat

Cutout in backing block holds slat.

ROUT

Frame

Add curved inserts to cut round slat bottoms.

75°

Screw frame to backing block.

A splined key joins the ends of Erpelding's chair-seat rim at the rear leg, above. The legs are fastened to the rim with socket-head cap screws threaded into propeller nuts mounted in a reinforcing block inside the rim. The curved clamping blocks of the chair-seat rim-bending form, below, which fit inside the laminated rim when it's being glued up, are shown resting on the metal pins that align the laminates.

In these knockdown spiral-stacking chairs, every piece was laminated and bent, then joined together with metal fasteners.

design was different enough to justify pursuing.

The final chair design evolved after many paper incarnations. I built a few models, but the mockups only frustrated my thinking. So I put the chair project on the back burner, and returned to it nearly a year later, visualizing a circular seat rim with legs and back supports coming off at points of a circumscribed hexagon. With an idea of what dimensions the chair would need to stack, I began building the bending forms. I guessed, and hoped the calculations were correct, but I could perfect the design only by building the chair.

I bent the backrest by clamping $\frac{1}{16}$-in. ash laminates in a cone-shaped form. The curve of the backrest tapers in radius from top to bottom to follow the bend in the upright back supports. Since the human back also tapers from shoulder to waist, this makes a comfortable chair.

The outside diameter of every seat rim had to be identical, or at least within a $\frac{1}{32}$-in. tolerance, or the chairs might not stack. The rim could have been glued up around a circular form and machined true later, but my shop isn't equipped to do this easily. Instead, I built a peripheral bending form, clamping up the seat rim inside it with curved blocks on the inside surface of the rim. Metal rods projecting from the bottom position the laminates in the same plane during glue-up. To ensure color and grain continuity in the set of chairs, I cut the outside laminate of each rim from the same piece of

wood. Planing each laminate before glue-up kept sanding to a minimum. I cut the laminates to a length just shy of the outside circumference. After removing the glued-up rim from the form, I slotted the break with a slotting cutter in a router, and fitted the joint with a key where it joins at the rear leg. I glued an additional reinforcing piece across the joint inside the rim. All of the rims have exactly the same outside diameter.

I drilled the rims and installed propeller nuts (available from Selby Furniture Hardware Co., 17 E. 22nd St., New York, N.Y. 10010) from the inside. Flat-head socket cap screws fasten the legs and back supports, which are also form-bent laminations, to the rim. I put an emery-cloth spacer between the rim and the legs to prevent slippage, and to serve as a shim so that the legs clear the seat rim of the chair underneath when the chairs are stacked.

There were few aesthetic decisions in this chair's design. Having stated the problem—a three-legged, knockdown, stacking chair—the design evolved in a purely functional and geometric way. Nearly all of the design decisions—the thickness and width of the members, the radii of the bends, the choice of fasteners and seat cushion—were arrived at as the simplest solution to the stated problem.

The third prototype, a leaning shelf system, was a reworking of an earlier design inspired by Italian designer Vico Ma-

A wedge joint holds the shelves together. The compound taper of the wedge in a dado forms a sliding dovetail and locks tight.

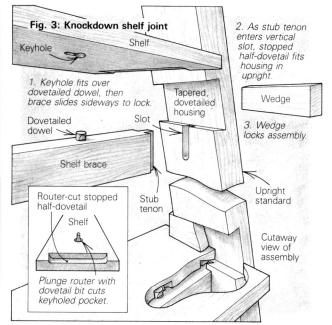

Knockdown leaning shelves are supported by both the wall and the floor. The horizontal braces under the shelves add strength and allow a longer shelf span.

gistretti. With minimal material and no bracing, the bookcase is very sturdy because weight is distributed between wall and floor. The design eliminates the need for shimming or fastening where floors slope, and the piece lends itself well to knockdown design.

I had originally intended to use metal hardware—a bolt passed through the upright standard and threaded into a nut embedded in the end of the shelf. While doodling, however, I came up with the wedge joint which, because it involves wood alone, fit my design philosophy more neatly. At first I used a wedge with a narrow tongue on the top edge which slid into a groove in the upright, but now I use a simpler joint. I've replaced the tongue and groove with a compound-taper wedge that acts like a sliding dovetail (figure 3).

To cut the edges of the dovetailed dadoes in the standards, I used a router with a 1-in. dia. dovetail bit, after wasting the material between with a ½-in. straight bit. A router-cut dovetail pocket on the underside of the shelf end fits into the dovetail-shaped bottom edge of the dado.

The horizontal braces under the shelves add strength and make a longer unsupported span possible (for ¾-in. shelving, 30 in. is about maximum before sagging occurs under load). The braces also stiffen the assembly against racking, act as a stop for books on the shelf below, and center each shelf with respect to the upright standards. The brace attaches underneath the shelf with a keyholed-pocket dovetail. Dovetailed dowels are driven into the edge of the brace, and dovetail slots are cut into the bottom of the shelf with a plunge router. The dowels enter the slot in the hole where the router bit was plunged in. The assembly locks tight when the brace is slid toward one end. A stub tenon on each end of the brace fits into a slot cut into the upright.

I am just now entering the third phase of the grant—researching the possibility of mass-producing the prototypes. Although I received only a one-year grant, I spent more than two years, on and off, working on the prototypes, and I am by no means finished with revisions.

To make an object is to arrange components into a new system. The assembly of parts can be final, the intersections fixed with glue, or it can be temporary—knockdown design. For me, the intersection of the components, the joinery, defines the form—not the other way around. Form is significant only insofar as it is incorporated into the structural integrity of the whole. The knockdown approach focuses on the relationships between the structure's components, and adds a new dimension to form—reconstructability. □

Curtis Erpelding designs and builds furniture in Seattle, Wash. His article about how to make slip joints using the radial-arm saw appears on page 39.

Pole-and-Wire Joinery
The quick way to build

by Len Brackett

Almost everyone at some time needs a temporary building or shelter, but most structures on the market are both expensive and time-consuming to erect, plus awkward to store once their purpose has been served. As a carpenter's apprentice in Japan, I encountered a method that is simplicity itself—lashing poles together with wire into a scaffolding that can support a roof. Because joints can be lashed easily and quickly anywhere along the poles, structures can be adapted to any site, even to rough terrain where prefabricated buildings cannot be used. In Japan the technique is used primarily for scaffolding, but also for wood-drying sheds, tool storage, and even for enclosing an entire temple while it is being worked on. I built my own 40-ft. by 60-ft. workshop this way; other applications are craft-fair booths, lean-tos, covered woodpiles and trellises, to name but a few.

Materials—Japan is blessed with some of the richest forests in the world, and wood has always been the traditional building material. Straw rope used to be the traditional binding for pole construction, but #9 annealed iron wire (not common baling wire), which neither rots nor frays and is stronger and faster to wrap than straw rope, is the material of choice today. You can buy it at most hardware stores or at building and agricultural-supply outfits. Enough wire to make one joint (about 3 ft. to join 5-in. dia. poles) will cost about 3ᶜ. Be sure to get annealed black wire; galvanized wire is too highly tempered, too brittle and too stiff to work.

As for poles, straight ones with little taper work best. They are lighter in proportion to their strength and easier to store, transport and position than ones with a lot of taper. Almost any species will work. We like to use Douglas fir or ponderosa pine because they are straight, strong and locally available. Poles larger than 8 in. in diameter are heavy and awkward. A butt diameter of 4 in. to 5 in. is best. All poles must be peeled. Bark holds moisture inside the poles, which fosters fungus growth and insect infestation (boring beetles especially), both of which can dangerously weaken a pole. Beware of rot, knots or other weaknesses, particularly if the pole is to bear horizontal loads. Poles should be stored in a dry place, vertically for maximum ventilation if stored outside.

The only tools you will need are a wire cutter (preferably the kind easily used with one hand) and a sturdy tapered spike for twisting the loops of wire. You could use an iron-worker's spud wrench, a large machinist's punch or even a very large nail.

The photos and sequence of drawings at right show how to wrap a joint and secure the wire knot. Practice a few times before you attempt to put up some structure—be sure you understand how the loop wraps around the tails, not vice-versa, or else the wire might fatigue and break off. Tighten the knot until the wire bites into the wood, then stop. Too much twisting will also weaken or break the wire. When

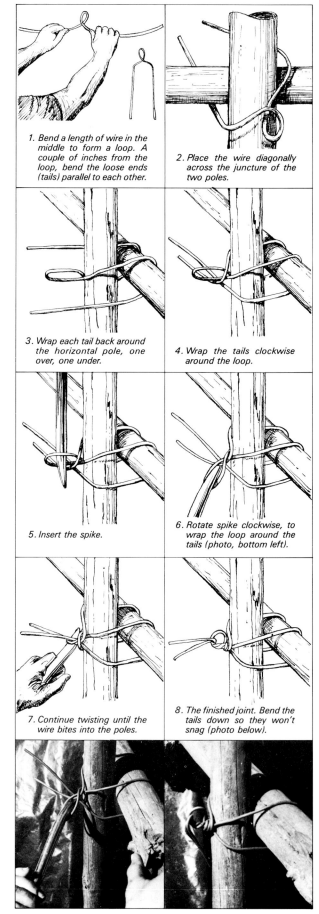

1. Bend a length of wire in the middle to form a loop. A couple of inches from the loop, bend the loose ends (tails) parallel to each other.

2. Place the wire diagonally across the juncture of the two poles.

3. Wrap each tail back around the horizontal pole, one over, one under.

4. Wrap the tails clockwise around the loop.

5. Insert the spike.

6. Rotate spike clockwise, to wrap the loop around the tails (photo, bottom left).

7. Continue twisting until the wire bites into the poles.

8. The finished joint. Bend the tails down so they won't snag (photo below).

Photos this page: Bob Ericson

Brackett's pole-and-wire shop is spacious and well ventilated; detail, left, shows cross-bracing at central post.

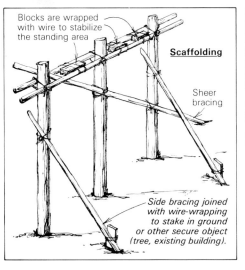

Blocks are wrapped with wire to stabilize the standing area

Scaffolding

Sheer bracing

Side bracing joined with wire-wrapping to stake in ground or other secure object (tree, existing building).

Ridgepoles

Rafter

Purlins

Roofing

Post

Nail

Pole building layout

The same basic layout can be used for the roof of a shed, lean-to, craft booth, etc.

done, bend the ends of the wire down to the wood, so they can't snag clothing or flesh.

With practice, you can make a joint in just a few seconds, especially if the wire is precut and hanging in your belt. The joints tend to self-tighten, but it's wise to check them periodically, especially if the poles are green or are exposed to humidity extremes. Poles shrink in dry weather, loosening the joint. You cannot tighten wire simply by pulling on a loose end, so wrap it tightly to start, as if it were wallpaper, smoothing out the wrinkles from the middle toward the ends. And of course, you should take the shortest and most direct route around the poles.

Building—Once you have mastered the way to twist the wire, you are ready to design a building. The following suggestions should help you get started. Pole-building layouts look a bit goofy, but are really quite rational, even if the rafters do hang from the purlins, ridgepole and top plates. Improvisation is the key. The structure must be kept rigid. Be sure to fix the poles to a stake in the ground or to some solid object to keep them from falling over as you build. Later they can be shear-braced with diagonal poles.

Pole-and-wire joinery is a very fast way to make scaffolding. Begin by fastening two horizontal poles at the required height, one on either side of the verticals. Jam spacer blocks of wood between the horizontal poles at 6-ft. intervals, then lash the horizontal poles tightly together, as shown in the drawing above left.

Pole buildings usually have corrugated steel roofing running parallel to the rafters. In this case, the purlins must be placed on top of the rafters and midspan between the top plates and the ridgepole (drawing above), and at the correct intervals for nailing the steel roofing to them. Rafters are fastened to the posts below and up against the top plate. The ridgepole can be placed on top of the central post for a gable roof; better yet, two ridgepoles can be fastened, one on either side of the central posts and high enough so the posts won't protrude through the roofing material. Put in the shear bracing, nail on the roofing and the roof is complete.

My own 2,400-sq. ft. shop was put together by three men in 2½ days, not including the time it took to peel the poles. It ought to last 15 to 20 years, and it cost about $600 for wire and roofing in 1976. Building inspectors have nightmares about such buildings, which are out of keeping with a heavily industrialized, consumer economy. But they do go together fast to make a variety of structures, they disassemble quickly and the materials remain available, unimpaired, for future use. In today's throw-away society, the ingenuity, simplicity and economy of pole buildings have great appeal. □

Len Brackett, of Nevada City, Calif., apprenticed for five years with temple builders in Kyoto, Japan.

Index

Acrylic adhesives, 102
Aliphatic resin glue (yellow), 97, 98
Allied International veneer, 52
Anaerobic adhesives, 102
Anderson, John F., on curved dovetails, 108-109
Annual rings, in Japanese frames, 21
Antique planes, 61
Apprenticeship system, Japanese, 18
Armstrong Products epoxy glues, 100, 101

Backsaws. See Dovetail saws.
Bairstow, John E.N., on decorative joinery, 82-85
Bandsaws:
    dovetails with, 112
    tenons on, 38
Bar clamps, 92
Barnsley, Edward, 49
Bed, platform knockdown, 114-115
Behlen hide glues, 98
Bertorelli, Paul, on dovetail jigs, 103-105
Biscuit joinery. See Plate joinery.
Bits, end-mill, for mortises, 47
Black and Decker dovetail jig, 103
Blood glue, 96
Boardman, Allan J., on precision, 2-5
Bosch Power Tool Corp.:
    address, 45
    dovetail jigs of, 103
    purchase of Stanley Tools by, 45
Bostik Thermogrip hot-melt glue, 101
Bowsaws, for tenons, 9
Box joints. See Finger joints.
Brackett, Len, on pole-wire joinery, 118-119
Breadboard ends, for doors, 69
Bridle joints:
    curved, 70, 73 (see also Mortise and tenon: curved.)
    disadvantages of, 6, 30-31, 63
    outlined decorative, 87
    on radial-arm saw (see Slip joints.)
    uses of, 7
Butler, J., panel-raising plane by, 61
Butterfly keys, 110-111
Butt joints, end-to-end, 67

Carcases, gluing-up of, 94, 95
    See also Frame and panel.
Carpenter's Wood Glue, 98
Casein glue, 96, 98
Chairs, three-legged stacking, 115-116
Chisels, for dovetails, 91
    See also Japanese chisels. Mortise chisels.
Circular-saw blades, for dovetails, 106-107
Clamping mortise and tenon, 28
Clamps, quick-release, 77
Classic Furniture Projects (Marlow), 89
Combination square, machinists', 3
Construction adhesive, 102
Contact cements, 101, 102
Craftsman. See Sears.
Cyanoacrylate glues, 101-102

Dadoes:
    Japanese, 22
    jig for, 5
Davies, Ben:
    on curved slot-mortise and tenon, 78-81
    on exterior doors, 56-59
DeStaco hold-down clamps, 77
Devcon glues, 100, 101
Domestic Architecture of Connecticut, The (Kelly), 63
Doors:
    battened, 68
    breadboard ends for, 69
    cleated and braced, 68
    cleated with sliding dovetails, 69
    coping on, 58-59
    design of, 58-59
    double-layered, 60
    dovetailed, 65
    exterior, 56-59
    hanging, 65
    with integral shelves, 69

lap-jointed, 68
layered, 68
louvered, making, 32-33
moldings for, 58-59
with mortise and tenon, 57
splined, 69
with steel rods, 69
with strap hinges, 68
stresses in, 57
    See also Shoji (Japanese sliding doors).
Dovetails:
    bandsawn, 112
    curved, layouts for, 109
    in door-frame, 65
    gluing up, 95
    with jigsaw, 91
    Keller jig for, 103-104
    Leigh jig for, 103, 104-105
    marking for, 90
    outlined decorative, 86
    pocket, with router, 115, 117
    on radial-arm saw, 39, 42-44
    repairing, 86
    routing, 90
    Sears jig for, 103, 104, 105
    on tablesaw, 106-107
    templates for, 89-91, 108-109
    three-way, 44
    See also Butterfly keys.
Dovetail saws, for tenons, 15
Dowels:
    for drawboring, 29
    in mortise and tenon, 57
Drawer pulls, routed inlaid, 74-75
Drilling through width of boards, 69
Drill presses:
    hollow chisel attachments for, 35-36
    mortising jigs for, 34-35, 36
Duginski, Mark, on tablesaw dovetails, 106-107
Duro Super Glue, 101

Ekstrom Carlson spiral end mill, 47
Elmer's glues, 97, 98, 101
Elu routers, 45
End-joining. See Butt joints.
Epoxy glues, 100
Erpelding, Curtis:
    on knockdown furniture, 114-117
    on slip joints with radial-arm saw, 39-44
    table by, 39
Evertite Glue, 98
Exterior doors, 56-59
    edge treatment of, 65

Fibonacci series explained, 59
Field, Isaac, molding plane by, 61
Fielded panel. See Frame and panel. Panels.
Fillers, cyanoacrylate glue formulations for, 102
Finger joints, decorative, 82-85
Fish glue, 96
Flounder Bay Boat Lumber, address, 100
Force Machinery Co., address, 111
Forms for chair seats, 116
Foxtail wedges. See Mortises: wedged, foxtail.
Frame and panel, 60
    in Arts and Crafts movement, 49
    basic, 48
    chamfered, 49-50
    defined, 48, 60
    and figured wood, 49
    gluing up, 94, 95
    making, with planes, 63
    plastic glass-retainer for, 58
    public demand for, 51
    routed profile on, 53
    variations with, 48-49
    See also Doors. Panels. Shoji (Japanese sliding doors).
Framing square, truing, 2
Franklin's glues, 97, 98
Frid, Tage:
    on bandsawn dovetails, 112
    on decorative joints, 86-88
    on door-making, 68-69
    on hanging doors, 65
    on mortise and tenon, 6-11
    on routing mortises, 45-47
Furniture, knockdown, designing, 114-117

Gimson, Ernest, 49
Glass in doors. See Exterior doors.
Glues:

rub joint with, 92-93
spreaders for, 92, 93
synthetic, compared, 100-102
types of, 96-99
Gluing up, 92-95, 99
    See also separate construction methods; separate joints.
Golden section and Fibonacci series, 59
Gougeon Brothers epoxy glues, 100, 101
Ground Hide Glue, 98

Hardware for furniture, 116
    See also Hinges.
Hide glues, 96, 98
Hinges:
    placing and inlaying of, 65
    routed inlaid, 76-77
    wooden, 113
Hold-downs, clamps for, 77
Holub, Dale, routed inlaid drawer pull by, 74-75
Hot-melt glues, 101, 102

Industrial Formulators epoxy glues, 100, 101

Japanese chisels, 24-25
Japanese houses, floor plan of, 19
Japanese planes, 20
Japanese saws:
    dovetail, for tenons, 15-16
    uses of, 22, 23, 25
Japanese sliding doors. See Shoji.
Japanese splitting gauge. See Splitting gauge, Japanese.
Japanese woodworking, methods of joinery in, 23-25
Jigs for precise handwork, 4-5
Joinery:
    decorative, 82-85
    designing, 27-29
    exposed, 44
    See also Japanese woodworking. Plate joinery.
Joints. See Bridle joints. Butt joints. Dovetails. Finger joints. Lap joints. Miters: splined. Mortise and tenon.
Jorgenson clamps, 92

Kahn, Louis, 44
Keller, David, dovetail jig by, 103-104
Kelly, J.F.: Domestic Architecture of Connecticut, The, 63
Kimball, S., panel-raising plane by, 61
Kirby, Ian J.:
    on frame-and-panel construction, 48-51
    on gluing up, 92-95
    on haunched mortise and tenon, 30-32
    on lock joints, 27-29
    on machine-made mortise and tenon, 34-38
    on mortise and tenon, 12-17
Kitchen cabinets, tambour doors for, 66-67
Klausz, Frank, on butterfly joint, 110-111
Knives for tenon marking, 14
Knockdown furniture. See Furniture, knockdown.
Krazy Glue, 101
Kumiko. See Shoji (Japanese sliding doors).

Lamello joining plate, 93
    See also Plate joinery.
Laminating, glues for, 100, 101
Lamination of chairs, 116
Lap joints, curved, 70, 73
Lego, William D., on radial-arm saw panels, 52-53
Legs:
    measuring for joints in, 95
    with rails, gluing up, 94, 95
Leigh Industries dovetail jig, 103, 104-105
Levels, truing, 67
Liquid Hide Glue, 98
Listel, 54, 55
Loctite Corporation anaerobic glues, 102
Louvers, router jig for, 32-33

Mackaness, Tim, on making folding screen, 113
Makita plunge-type routers, 45-46, 47
Mallets, for mortising, 14

Mantels, 60, 63
Marking for precision, 3
Marlow, A.W.: Classic Furniture Projects, 89
McQuilkin, Bruce, 74
    routed inlaid hinge by, 76-77
Measuring:
    legs, sequence for, 95
    for precision, 3
    squareness, with sticks, 94, 95
    with sticks, 21, 22, 23
Miters:
    dovetail-keyed, 83
    masons', 49, 50
    splined, decorative, 87-88
    on stuck moldings, 63-64
    tolerance in, 2-3
    trimming of, block for, 64
Molding planes, 61, 63
Moldings:
    on doors, 58-59
    scribed (see Scribed joints.)
    stuck, 60 (see also Frame and panel.)
    vocabulary of, 54
    See also Frame and panel. Molding planes. Panels.
Mortise and tenon:
    angled, 7, 11
    blind-doweled, 57
    clamping, 17
    coped, advantages of, 64
    curved, 78-81
    designing, 12-13
    draw-bored, 27, 29
    with folding wedges, 29
    gluing, 11, 93-95
    haunched, 30, 31-32, 49
    Japanese, 20-25
    by machine, 34-38
    making, 6-11
    mitered, 8, 64
    pinned, 56-57
    with round corners, 7
    slip-joint, 39-44
    tools for, 13-14
    tusked, 27, 29
    wedged, foxtail, 28
    wedged through, 27-28, 57
    See also Mortises. Tenons.
Mortise chisels, 12-17
Mortise gauge, 13-14
Mortisers, 35-37
Mortises:
    angled, jig for, 46, 47
    bits for, 35
    chiseling, 10, 13-15
    curved, 78-81
    in curved pieces, 47
    drilling, 10, 15, 34, 35
    drilling jigs for, 34, 35, 36
    haunched, 6, 31, 32
    hollow-chisel attachment for, 35-36
    hollow-chiseling, 36-37
    Japanese, 24-25
    mitered haunched, 6
    plunge-type router jigs for, 45-47
    rasp with, 31-32
    router bits for, 47
    trying, 16
    See also Mortise and tenon. Mortise gauge. Tenons.
Mortising planes, for door hanging, 65
Mullion, defined, 60
    See also Panels.
Muntin, 60
    See also Panels.
Mustoe, George, on glues, 96-99, 100-102

Nakashima, George, 110
National Casein glues, 98
New Departures router cutters, 71
Nuts, propeller, source for, 116
Nylin, Judy, tapestry by, 113

Odate, Toshio, on Japanese sliding doors (shoji), 18-26
Ogee, 60
Okie, Grif, on routers, 74-77
Onsrud Cutter spiral end mill, 47
Ovolo, 54, 55, 60

Panel-raising planes. See Raising planes.
Panels, 60
    authentic reproduction of, 63
    with feathered edge, advantage of, 60
    fielded, 50-51
    making, sequence of, 63
    overlapped, 50, 51

profiles of, 60
rabbetted and beaded, 51
raised, 48, 50-51, 53, 62, 63
raised, radial-arm saw jig for, 52-53
with tongue and groove, 60
unfielded, 50
variations with, 48-49
vocabulary of, 60
in walls, 63
See also Frame and panel.
Pearl Hide Glue, 98
Phenol-formaldehyde glue, 97, 98
Pin routers, U-shaped finger-joint jigs for, 84-85
Pins. *See* Dovetails.
Plastic resin glues, 98
Plate joinery:
   advantages of, 93
   on radial-arm saw, 67
Plow planes, 61, 63
Pole-and-wire joinery, 118-119
Polyester resins for fiberglassing, 100-101
Polyvinyl resin glue (white), 97, 98
Porter Cable dovetail jig, 103
Precision in woodworking, 2-5
Protein glues, 96
PVA glues. *See* Aliphatic resin glue (yellow). Polyvinyl resin glue (white).

Radial-arm saw:
   accuracy of, 39
   bridle joints on (*see* Radial-arm saw: slip joints on.)
   dovetails on, 42-44
   plate joinery with, 67
   raised-panel jig for, 52-53
   slip joints on, 39-42
   tenons on, 38, 40-41
Rails. *See* Frame and panel. Panels.
Raising planes, 60-63
   antique, 61
RC-76 glue, 97, 98
Record clamps, 92
Resorcinol-formaldehyde glues, 98
Reynolds, William F., on louvered doors, 32-33
Riordan, Charles F., on template dovetails, 89-91
Roberts, Kenneth D.:
   raising planes of, 61
   Wooden Planes in 19th Century America, 61
Routers:
   ball-bearing cutters for, 70-72
   butterfly mortises with, 111
   choosing, 45-46
   curved joinery with, 70-73
   for dovetails, blind, 90
   dovetail jigs for, commercial, 103-105
   drawer pulls with, 74-75
   guide bushings for, 80
   hinges with, inlaid, 76-77
   louvered doors with, jig for, 32-33
   for mortise and tenon, 38
   for mortise and tenon, curved, 78-81
   mortise cutters for, 47
   mortise fences for, 38
   mortise jigs for, 46-47, 76
   plunge-type, mortising with, 45-47
   pocket dovetail jig for, 115, 117
   templates for, curved, 70, 72
   tenon jig for, 37

Sanding, tapered discs for, 53
Scratch gauges, machinists', 89
Screens, wooden-hinged, 113
Screws, flat-head socket cap, 116
Scribed joints, 54, 55, 64
Sears:
   dovetail jig, 103, 104, 105
   radial-arm saw, 10-in. Craftsman, 39, 40, 43
   sanding discs, tapered, 53
Selby Furniture Hardware, 116
Shelves, wall-leaning knockdown, 117
Sheppard, Morris J., on scribed joints, 54-55
Shoji (Japanese sliding doors), 18-26
Shooting board, 4
Shoulder planes:
   for precise cuts, 4
   on tenons, 15
Sig Model Airplane Epoxy, 100
Slip joints, 39-44
Slot mortisers. *See* Mortisers.

Slot mortises. *See* Mortises: curved.
Smith, A., panel-raising plane by, 61
Soybean glue, 96
Spangler, Ned, louvered doors by, 32-33
Spline miter. *See* Miters: splined.
Splitting gauge, Japanese, 20
Spring, of planes, defined, 60
Squares. *See* Try squares.
Stanley plunge-type routers No. 90105 and No. 90303, 45
Starr, Richard, on tambour kitchen cabinets, 66-67
Starrett Tools, scratch gauge of, 89
Steere, W.B., 61
Sticking, 54, 55, 60, 63
   See also Frame and panel.
Sticks. *See* Measuring: squareness, with sticks; with sticks. Winding sticks.
Stiles:
   annual ring orientation for, 21
   defined, 60
   See also Frame and panel.
Surfacing for gluing up, 99
Sweeney, Jim:
   chest-of-drawers by, 70, 71
   on curved joinery, 70-73
Swingline Fix Stix hot-melt glue, 101

Tablesaws:
   dovetails on, 106-107
   tenons on, 37, 38
Tambours, 66-67
Templates:
   for dovetails, 89
   for scribed joints, 54, 55
Tenons:
   chiseling of, 15-16
   correcting precisely, 3-4
   curved, 78-81
   drawboring of, 64
   Japanese chamfering of, 22
   narrow, 7
   on radial-arm saw, 40-41
   repairing, 8, 11
   router jig for, 37
   sawing, 9, 10, 11, 15
   saws for, 15-16
   shoulder for, 6-7, 14, 38
   tablesaw jig for, 37
   through-wedged, 44
   trying, 16-17
   wedged, 7
   wedged, for doors, 57
   wide, 7
   See also Mortise and tenon.
Thompson, Mousey, panel texture of, 49
Titebond Glue, 97, 98
TRW router cutters, 71
Try squares, testing and truing, 2-3

Urea-formaldehyde glues, 97-98
U.S. Plywood Resorcinol Waterproof Glue, 98

Vandal, Normal L., on paneling with planes, 60-64
Velpec router cutters, 111
Veneer, thermosetting-adhesive backed, 52

Waals, Peter, 49
Warner, Patrick, on decorative frame-and-panel, 53
Waterproof glues, 98
Weed, Hazel, 66
Weed, Walker, tambour doors of, 66-67
Weldbond glues, 97, 98
Wetzler Clamp Co., 92
White glues, 97, 98
Wilhold glues, 97, 98
Winding sticks, 3
Wooden Planes in 19th Century America (Roberts), 61
Woods, J., combination plane by, 61
Workbenches, Japanese tripod, 20

Yellow glues, 97, 98

Zip Grip 10 glue, 101

Fine WoodWorking

To subscribe

If you enjoyed this book, you'll enjoy *Fine Woodworking* magazine.
Use this card to subscribe.

1 year (6 issues) for just $16—$5 off the newsstand price.

Canadian subscriptions: $19/year; other foreign: $20/year. (U.S. funds, please)

Name _____

Address _____

City _____ State _____ Zip _____

☐ My payment is enclosed. ☐ Please bill me.

☐ Please send me more information about Taunton Press Books.

☐ Please send me information about *Fine Woodworking* videotapes.

FPCT

Fine WoodWorking

To subscribe

If you enjoyed this book, you'll enjoy *Fine Woodworking* magazine.
Use this card to subscribe.

1 year (6 issues) for just $16—$5 off the newsstand price.

Canadian subscriptions: $19/year; other foreign: $20/year. (U.S. funds, please)

Name _____

Address _____

City _____ State _____ Zip _____

☐ My payment is enclosed. ☐ Please bill me.

☐ Please send me more information about Taunton Press Books.

☐ Please send me information about *Fine Woodworking* videotapes.

NO POSTAGE
NECESSARY
IF MAILED
IN THE
UNITED STATES

BUSINESS REPLY CARD
FIRST CLASS PERMIT No. 19 NEWTOWN, CT

POSTAGE WILL BE PAID BY ADDRESSEE

The Taunton Press
52 Church Hill Road
Box 355
Newtown, CT 06470

NO POSTAGE
NECESSARY
IF MAILED
IN THE
UNITED STATES

BUSINESS REPLY CARD
FIRST CLASS PERMIT No. 19 NEWTOWN, CT

POSTAGE WILL BE PAID BY ADDRESSEE

The Taunton Press
52 Church Hill Road
Box 355
Newtown, CT 06470